Protein Hydrolysates in Biotechnology

Vijai K. Pasupuleti · Arnold L. Demain
Editors

Protein Hydrolysates in Biotechnology

Springer

Editors
Vijai K. Pasupuleti
SAI International, Inc.
Geneva, IL
USA
vijai@saiintl.com

Arnold L. Demain
Research Institute for Scientists Emeriti
(R.I.S.E.)
Drew University
Madison, NJ
USA
ademain@drew.edu

DOI 10.1007/978-1-4020-6674-0
Springer Dordrecht Heidelberg London New York

Printed on acid-free paper

Springer is part of Springer Science+Business Media (www.springer.com/mycopy)

Preface

Protein hydrolysates, otherwise commonly known as peptones or peptides, are used in a wide variety of products in fermentation and biotechnology industries. The term "peptone" was first introduced in 1880 by Nagelli for growing bacterial cultures. However, later it was discovered that peptones derived from the partial digestion of proteins would furnish organic nitrogen in readily available form. Ever since, peptones, which are commonly known as protein hydrolysates, have been used not only for growth of microbial cultures, but also as nitrogen source in commercial fermentations using animal cells and recombinant microorganisms for the production of value added products such as therapeutic proteins, hormones, vaccines, etc.

Today, the characterization, screening and manufacturing of protein hydrolysates has become more sophisticated, with the introduction of reliable analytical instrumentation, high throughput screening techniques coupled with statistical design approaches, novel enzymes and efficient downstream processing equipment. This has enabled the introduction of custom-built products for specialized applications in diverse fields of fermentation and biotechnology, such as the following.

1. Protein hydrolysates are used as much more than a simple nitrogen source. For example, the productivities of several therapeutic drugs made by animal cells and recombinant microorganisms have been markedly increased by use of protein hydrolysates. This is extremely important when capacities are limited.
2. Protein hydrolysates are employed in the manufacturing of vaccines by fermentation processes and also used as vaccine stabilizers.
3. Protein hydrolysates are being used in large-scale industrial fermentations as sources of nitrogen and unknown growth factors, such as certain peptides, etc. They are also useful in diagnostic media to grow microorganisms in Petri plates and to detect pathogens and perform antibiotic sensitivity tests.
4. Protein hydrolysates are used in regular diets as well as prescription diets for companion animals.
5. Protein hydrolysates play an important role in animal nutrition, especially for raising healthy animals with increased immune resistance.
6. Protein hydrolysates are used as plant growth regulators to increase commercial crop yields as well as to control weeds in plants.

Thus protein hydrolysates industry is growing rapidly with wide applications in biotechnology, yet there has been no single book that describes (i) the challenges and opportunities for manufacturers and end users, (ii) the techniques used in manufacturing, characterization and screening of protein hydrolysates, and (iii) the applications of protein hydrolysates in a wide variety of industries, e.g., that of fermentation for production of many primary and secondary metabolites of microorganisms, and in the rapidly growing area of industrial biotechnology for production of biopharmaceuticals. One of the misconceptions involving the use of protein hydrolysates in fermentations is that they are being used merely as a nitrogen source. However, the functionality of the product obtained is not necessarily due solely to protein hydrolysates in general, i.e., it may be due to specific peptides, or the combination of peptides, or to non-protein components in the product. This is due to the fact that the preparations may contain carbohydrates, lipids, micronutrients, etc. Indeed, some manufacturers deliberately add such factors into the process to bring about unique functionality. Since only a handful of manufacturers dominate the market, this tends to keep the manufacturing process proprietary, making it harder to understand and appreciate its fine points. This book will close the gap by revealing valuable information on the latest developments in this vital and tremendously important field.

Vijai K. Pasupuelti
SAI International, 1436 Fargo Blvd.,
Geneva, IL 60134,
USA

Arnold L. Demain
Research Institute for Scientists Emeriti (R.I.S.E),
Drew University, NJ 07940,
USA

Acknowledgements

Our special thanks are due to all the contributors; without their dedication, hard work, patience and time to share their expertise; this book would not have been possible.

We also thank Jacco Flipsen for coordinating the book and Ineke Ravesloot for editorial assistance from Springer publishers.

VKP likes to thank his father P.V. Subba Rao; wife Anita; and sons, Anoop and Ajai, for providing support, energy and enthusiasm in completing this book.

ALD appreciates the patience of his wife JoAnna for all those hours he spent on the computer and his daughter, Pamela, son, Jeffrey, and colleague Vince Gullo for their assistance in solving computer problems. He also expresses thanks to coauthor Vijai for inviting him to participate in the editing of this fine book.

Contents

Contributors

James Babcock
Sheffield Bio-Science, 283 Langmuir Lab, 95 Brown Road, Ithaca, NY 14850, USA

Steven Braun
GR&D Operations, Global Research and Development, Mead Johnson,
2400 W. Lloyd Expressway, Evansville, IN 47721, USA

Robert Chang
FibroGen, Inc., 409 Illinois Street, San Francisco, CA 94158, USA

Nick Christians
Department of Horticulture, Iowa State University, 133 Horticulture Building,
Ames, IA 50011, USA

Mirjana Curic-Bawden
ITC Dairy, Chris Hansen Inc., Milwaukee, WI, USA

Arnold L. Demain
Research Institute for Scientists Emeriti (R.I.S.E), Drew University,
NJ 07940, USA

František Franek
Laboratory of Growth Regulators, Institute of Experimental Botany,
Radiová 1, CZ-10227 Prague 10, Czech Republic

Chris Holmes
Eli Lilly and Company, Indianapolis, IN 46285, USA

Hans Huttinga
Sheffield Bio-Science, Veluwezoom 62, 1327 AH Almere, The Netherlands

David Jayme
GIBCO Cell Culture, Invitrogen Corporation, Grand Island, NY, USA
Brigham Young University, 55-220 Kulanui Street, Laie, HI, USA

Dianna Liu
Nee Extract Private Limited, Brisbane, Queensland, Australia

Juliet Lobo-Alfonso
GIBCO Cell Culture, Invitrogen Corporation, Grand Island, NY, USA

Eric Lohry
Nutra-Flo Protein and Biotech Products, Sioux City, IA 51106, USA

James C. Matthews
Department of Animal Science, University of Kentucky, Lexington, KY, USA

Jeff McCalla
Nutra-Flo Protein and Biotech Products, Sioux City, IA 51106, USA

R. Meganathan
Department of Biological Sciences, Northern Illinois University, DeKalb, IL 60115, USA

Tilak W. Nagodawithana
Esteekay Associates, Inc., Milwaukee, WI 53217, USA

Lynn Nelles
Kemin Industries, Inc., 2100 Maury Street, Des Moines, IA 50317, USA

David Olsen
FibroGen, Inc., 409 Illinois Street, San Francisco, CA 94158, USA

Vijai K. Pasupuleti
SAI International, Geneva, IL 60134, USA

Shifa Patel
Department of Biological Sciences, Northern Illinois University, DeKalb, IL 60115, USA

James W. Polarek
FibroGen, Inc., 409 Illinois Street, San Francisco, CA 94158, USA

Paul Price
GIBCO Cell Culture, Invitrogen Corporation, Grand Island, NY, USA

Yamini Ranganathan
Department of Biological Sciences, Northern Illinois University, DeKalb, IL 60115, USA

André Siemensma
Sheffield Bio-Science, Veluwezoom 62, 1327 AH Almere, The Netherlands

Nayan B. Trivedi
Trivedi Consulting, Inc., Princeton, NJ 08540, USA

Madhavi (Soni) Ummadi
Technical Applications Group, Dreyer's Grand Ice Cream, Bakersfield, CA, USA

Jay Bryan Unruh
Department of Environmental Horticulture, West Florida Research and Education Center, IFAS, University of Florida, Gainesville, FL 32565, USA

Terry Waugh
Nutra-Flo Protein and Biotech Products, Sioux City, IA 51106, USA

Chris Wilcox
Sheffield Bio-Science, 3400 Millington Road, Beloit, WI 53511, USA

Kim E. Williams
FibroGen, Inc., 409 Illinois Street, San Francisco, CA 94158, USA

Brian M. Zhanghi
Department of Animal Science, University of Kentucky, Lexington, KY, USA

Chapter 1
Applications of Protein Hydrolysates in Biotechnology

Vijai K. Pasupuleti, Chris Holmes, and Arnold L. Demain

Abstract By definition, protein hydrolysates are the products that are obtained after the hydrolysis of proteins and this can be achieved by enzymes, acid or alkali. This broad definition encompasses all the products of protein hydrolysis – peptides, amino acids and minerals present in the protein and acid/alkali used to adjust pH (Pasupuleti 2006). Protein hydrolysates contain variable side chains depending on the enzymes used. These side chains could be carboxyl, amino, imidazole, sulfhydryl, etc. and they may exert specific physiological roles in animal, microbial, insect and plant cells. This introductory chapter reviews the applications of protein hydrolysates in biotechnology. The word biotechnology is so broad and for the purpose of this book, we define it as a set of technologies such as cell culture technology, bioprocessing technology that includes fermentations, genetic engineering technology, microbiology, and so on. This chapter provides introduction and leads to other chapters on manufacturing and applications of protein hydrolysates in biotechnology.

Keywords Protein hydrolysates • Biotechnology • Applications • Cell culture

Introduction

Protein hydrolysates, commonly known as peptones or peptides, are used in a wide variety of products in the fermentation and biotechnology industries (Pasupuleti 2006). The art of manufacturing and application of protein hydrolysates goes back

V.K. Pasupuleti (✉)
SAI International, Geneva, IL 60134, USA
e-mail: Vijai1436@sbcglobal.net

C. Holmes
Eli Lilly and Company, Indianapolis, IN 46285, USA

A.L. Demain
Research Institute for Scientists Emeriti (R.I.S.E), Drew University, NJ 07940, USA

V.K. Pasupuleti and A.L. Demain (eds.), *Protein Hydrolysates in Biotechnology*,
DOI 10.1007/978-1-4020-6674-0_1,

to the days when meat hydrolysates began their use to grow microbial cells (Nagelli in general, i 1880). Today, the manufacturing of protein hydrolysates has become more sophisticated with the introduction of novel enzymes and efficient downstream processing equipment to custom-build products for specialized applications in biotechnology (Heidemann et al. 2000).

With the increased knowledge, understanding and sophistication of manufacturing of protein hydrolysates has led to their applications in areas of fermentation and biotechnology such as medicine, agriculture, industrial fermentations, production of recombinant proteins, diagnostic media, bioremediation, weed control for plants and for healthy growth of young animals and companion animals. Throughout this book, authors have discussed these specialized applications of protein hydrolysates.

Protein hydrolysates are particularly useful in the biotechnological production of monoclonal antibodies (mabs), peptides and therapeutic proteins.

1. The contribution of protein hydrolysates is much more than that of supplying simple nitrogen sources. They are useful in increasing monoclonal antibody production by some unknown mechanism, and also for increasing the productivities of several therapeutic drugs made by animal cells and recombinant microorganisms. Chu and Robinson (2001) have reviewed the FDA biological license approvals from 1996 to 2000 and found that 21 out of 33 were manufactured using mammalian cell culture. The media of one third of these have not been revealed and only one company claimed to use protein hydrolysates in the medium. Anecdotal evidence suggests that more manufacturers use protein hydrolysates in their media and keep it as a trade secret. Some manufacturers file the patents and a partial list of patents is given in the reference section (World Patents). A sampling of published US Patents/Applications by major biopharmaceutical companies using protein hydrolysates by a variety of cell lines is discussed in Chapter 3.

   Protein hydrolysates are also used in the manufacture of vaccines and as an adjuvant in vaccines as seen in Table 1.

   Chapter 2 reviews in detail with the state of the art of manufacturing of protein hydrolysates for applications in biotechnology. The role of protein hydrolysates, their benefits and limitations in animal cell culture are covered in detail in Chapters 3–5.
2. Protein hydrolysates are widely used in the manufacture of probiotics, starter cultures and fermented products. Sobharani and Agrawal (2009) have demonstrated that the viability of probiotic culture was enhanced with supplementation of adjuvants like tryptone, casein hydrolysate, cysteine hydrochloride and ascorbic acid. Chapter 6 reviews the use of protein hydrolysates in industrial starter culture fermentations.
3. In industrial fermentations, protein hydrolysates have been used for several decades to supply nitrogenous compounds such as peptides, amino acids and to increase productivities and yields. They are also employed in diagnostic media to grow micro-organisms (BD Bionutrients Technical Manual 2006). Chapter 7 reviews replacement of tryptone in microbiological media by protein hydrolysates from non-bovine and plant sources.

Table 1 Excipients included in US vaccines. Includes vaccine ingredients (e.g., adjuvants and preservatives, protein hydrolysates are highlighted) as well as substances used during the manufacturing process, including vaccine-production media that are removed from the final product and present only in trace quantities

Vaccine	Contains
DTaP (Daptacel)	Aluminum phosphate, ammonium sulfate, **casamino acid**, dimethyl-betacyclodextrin, formaldehyde or formalin, glutaraldehyde, 2-phenoxyethanol
DTaP-IPV (Kinrix)	Aluminum hydroxide, bovine extract, formaldehyde, **lactalbumin hydrolysate**, monkey kidney tissue, neomycin sulfate, polymyxin B, polysorbate 80
DTaP-HepB-IPV (Pediarix)	Aluminum hydroxide, aluminum phosphate, bovine protein, **lactalbumin hydrolysate**, formaldehyde or formalin, glutaraldhyde, monkey kidney tissue, neomycin, 2-phenoxyethanol, polymyxin B, polysorbate 80, yeast protein
Hib/Hep B (Comvax)	Amino acids, aluminum hydroxyphosphate sulfate, dextrose, formaldehyde or formalin, mineral salts, sodium borate, **soy peptone**, yeast protein
Hep B (Recombivax)	Aluminum hydroxyphosphate sulfate, amino acids, dextrose, formaldehyde or formalin, mineral salts, potassium aluminum sulfate, **soy peptone**, yeast protein
Pneumococcal (Prevnar)	Aluminum phosphate, amino acid, **soy peptone**, yeast extract
Zoster (Zostavax)	Bovine calf serum, **hydrolyzed porcine gelatin**, monosodium L-glutamate, MRC-5 DNA and cellular protein, neomycin, potassium phosphate monobasic, potassium chloride, sodium phosphate dibasic, sucrose

Downloaded on June 16, 2009 from http://www.cdc.gov/vaccines/pubs/pinkbook/downloads/appendices/B/excipient-table-2.pdf

4. Special applications of protein hydrolysates include use as plant growth regulators and to increase pest resistance to plants (Inagrosa 2002; Figueroa et al. 2008). Protein hydrolysates are used in bioremediation, especially to boost the growth of microorganisms. Chapter 8 reviews the use of protein hydrolysates for weed control.
5. Chapter 9 provides a detailed review of the physiological importance and mechanisms of protein hydrolysate absorption. For animal and pet nutrition, protein hydrolysate are being used in regular as well as prescription diets of companion animals. Chapters 10 and 11 review in detail the applications of protein hydrolysates in animal nutrition and companion animals.

Finally, extending the theme of non-animal protein hydrolysates, Chapter 12 discusses the development of novel recombinant human gelatins as replacements for animal-derived gelatin hydrolysates in pharmaceutical applications.

What Are Protein Hydrolysates?

By definition, protein hydrolysates are the products obtained after the hydrolysis of proteins as achieved by acid, alkali, enzymes and fermentation methods. This broad definition encompasses all of the products of protein hydrolysis, i.e., peptides,

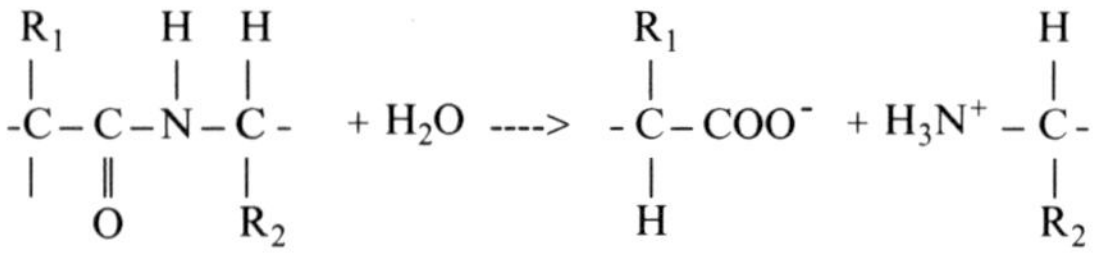

Adapted from Jens Adler-Nissen (1986)

amino acids, minerals, carbohydrates and lipids if they are present in the substrate or enzyme (sometimes animal tissues are used as an enzyme source) along with the protein.

The protein hydrolysates are typically characterized by the degree of hydrolysis, i.e., the extent to which the protein is hydrolyzed. The degree of hydrolysis (DH) is measured by the number of peptide bonds cleaved, divided by the total number of peptide bonds and multiplied by 100. The number of peptide bonds cleaved is measured by different test methods and the most popular are: Formol titration (Sorensen, 1907) or the OPA method (Church et al., 1985) or the trinitrobenzene sulphonic acid, TNBS (Fields 1971) reagent method, and reported as AN (alpha amino nitrogen). The Kjeldahl, or Dumas direct combustion methods, are typically used to measure TN (total nitrogen).

$$\%DH = \frac{\text{AN of protein hydrolysate} - \text{AN of protein} \times 100}{\text{Total Nitrogen of the protein}}$$

The common practice in industry is to measure AN and TN and multiply by 100 to report the AN/TN ratio which is nothing more than the degree of hydrolysis of the protein.

Protein hydrolysates are commonly known as peptones, protein fission products, peptides and hydrolyzed proteins; (Acid-HP, hydrolyzed acid protein and HVP, hydrolyzed vegetable protein). Typically, Acid-HPs and HVPs are used in the food industries and not in industrial fermentations and animal cell culture because of very high ash and low protein content.

Evolution of Protein Hydrolysates

The first application of protein hydrolysates dates back to 1880 when Nagelli used it to cultivate microbes. The next recorded use was around 1900 for preparation of Voges-Proskauer and McConkey media. In 1914, Difco Laboratories introduced Bacto-Peptone for use in bacteriological media. Since then, the uses of protein hydrolysates have become widely popular for the growth of microorganisms (Fig. 1).

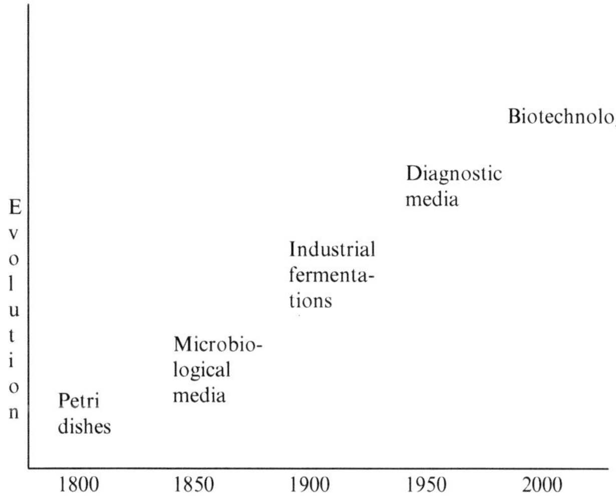

Fig. 1 Applications of protein hydrolysates over the years

Applications of Protein Hydrolysates in the Biotechnology Industry

The word "Biotechnology" is very broad and has been freely and widely used to define a wide variety of topics but here, for the purpose of this book, we define biotechnology in a broader sense that encompasses molecular biology, microbiology, fermentations, animal, insect and plant cell culture, and young animals and pets where protein hydrolysates are used as a nitrogen and nutritional source. Applications of protein hydrolysates in biotechnology are briefly discussed leading to other chapters where they are discussed in greater detail.

1. *Protein production in microbial cells. Escherichia coli* is the most popular bacterium for production of recombinant proteins. The traditional media for this fermentation process contain tryptone (Franchon et al. 2006). With respect to yeasts, *Pichia pastoris* is widely used and its media contain peptone which not only supplies nitrogen, but also reduces protein degradation by acting as an excess substrate for harmful proteases (Cregg et al. 1993). Another yeast often used is *Saccharomyces cerevisiae* and, in this case, peptone is also the usual nitrogen source (Penttila et al. 1987). With filamentous fungi such as *Aspergillus awamori*, again peptone is the favored nitrogen source (Lamsa and Bloebaum 1990).
2. *Protein production in mammalian cells.* Production of proteins, including monoclonal antibodies, in mammalian cells is usually carried out in Chinese Hamster Ovary (CHO) cells or non-secreting mouse myeloma (NSO) cells using

media containing animal serum. However, it is interesting to note that as early as Taylor et al. (1972 and 1974) studied a serum substitute for mammalian cells in culture using peptones. Until the Bovine Spongiform Encephalopathy (BSE) issue have became serious, serum was the favorite but after that, the trend has shifted towards serum-free media. This situation in which humanized antibodies are now produced in fungal cells such as *P. pastoris* (Kunert et al. 2008) or *Aspergillus niger* (Ward et al. 2004). This trend could see expanded use of protein hydrolysates, especially those from vegetable proteins like soy, wheat and pea in the future.

3. *Protein production in insect cell cultures.* The baculovirus expression vector system is the common process for producing proteins using insect cells. Such a system using *Spodoptera frugiperda* (Sf-9) cells employs a medium that contains lactalbumin hydrolysate (Neutra et al. 1992). Most insect cell media contain fetal bovine serum which is the most expensive component of the medium, is variable from lot-to-lot, may contain cytotoxic materials, is susceptible to contamination by viruses and mycoplasma, and causes problems such as foaming and difficult product purification. Furthermore, serum is not needed for insect cell growth and much effort has been put into serum-free media which often contain peptone or lactalbumin hydrolysate (Agathos et al. 1990).
4. *Vaccines.* The use of bovine serum to make certain vaccines has been looked down upon because of a possible adverse effect of prion diseases such as Mad Cow Disease (BSE). In such a disease, prions infect neurons of the brain. Thus, vegetable protein hydrolysates can be very useful for replacement of animal protein hydrolysates in preparation of media for production of toxins to be used to produce animal-free vaccines. For example, in the preparation of tetanus toxin, which is used to make the toxoid for immunization, growth of *Clostridium tetani* traditionally is done in media containing animal and dairy products (e.g., meat extracts, brain heart infusion, casein hydrolysates). As a result, toxoids often contain undesirable formalin adducts of animal proteins. To avoid this problem, a new medium containing hydrolyzed soy proteins has been devised which yields even higher titers of tetanus toxin than the old traditional medium (Demain et al. 2005, 2007; Fang et al. 2006).
5. *Plant cell culture.* The most important product made in plant cell culture is Taxol (paclitaxel), one of the most successful anti-cancer agents known. It is produced in plant cell suspension culture by *Taxus chinensis* and *Taxus yunnanensis*. The standard medium for the former culture contains casein hydrolysate (Choi et al. 2000) while that of the latter includes lactalbumin hydrolysate (Zhang et al. 2002).
6. *Production of primary metabolites.* One of the major contributions of microorganisms in industry is the production of primary metabolites such as amino acids, vitamins, flavor nucleotides, etc. Many of these fermentation processes utilize protein hydrolysates. For example, manufacture of L-glutamic acid is the major primary metabolite made in industry (1.5 million tons per year). *Corynebacterium glutamicum* is one of the cultures used to produce glutamic acid commercially in a medium containing soybean protein hydrolysate (Kataoka et al. 2006).

Future Directions

The most important development that has to take place is the partnership between protein hydrolysate manufacturers and the end users. This partnership will enable understanding of the capabilities of manufacturers and the requirements of end users which may lead to more defined and better products.

Genetic engineering might play a role in developing animal-free and defined products. One of the recent developments is to produce protein hydrolysates (peptides) by fermentation. Olsen et al. (2010) describes the whole process in detail in Chapter 12.

Major companies like Ajinomoto, Kyowa Hakko and others are manufacturing peptides by fermentation for use in cell culture and industrial fermentations. We hope that in the future, more products and processes like this will come into play as they not only alleviate the BSE problem but also provide consistency from batch to batch and provide higher confidence levels similar to defined media components.

Sophisticated analytical techniques such as NMR, MALDI/TOF, peptide mapping are being applied to characterize (fingerprint) protein hydrolysates and this coupled with spent media analysis may give better insight. This knowledge could be potentially utilized in the manufacturing of protein hydrolysates to tailor-make products for specific applications and to avoid variations from batch to batch.

Manufacturers, by using standardized units of enzyme, instituting in-process controls utilizing membrane technologies for purification, and establishing functional specifications, will get closer to producing consistent products.

One could foresee a major development in the enzymes used for manufacturing protein hydrolysates.

The design of experiments using factorial or fractional factorial methods and high throughput screening enables the end user to screen a variety of protein hydrolysates and dose response. Perhaps a combination of protein hydrolysates would be better in some instances.

We are optimistic and believe that in the near future all of the above technologies and partnerships between the manufacturers and end-users will bring exciting new developments in the applications of protein hydrolysates in biotechnology.

References

Agathos SN, Jeong Y-H, Venkat K (1990) Growth kinetics of free and immobilized insect cell culture. Biochemical Engineering VI. Ann NY Acad Sci 589:372–398

BD Bionutrients Technical Manual (2006) 3rd edn. Revised October 2006

Choi H-K, Kim S-I, Son J-S, Hong S-S, Lee H-S, Lee H-J (2000) Enhancement of paclitaxel production by temperature shift in suspension culture of *Taxus chinensis*. Enzyme Microb Tech 27:593–598

Chu L, Robinson D (2001) Industrial choice for protein production by large scale cell cultures. Curr Opin Biotechnol 12:180–187

Church FC, Porter DH, Catignani GL, Swaisgood HE (1985) An *o*-phthalaldehyde spectrophotometric assay for proteinases. Anal Biochem 146:343–348

Cregg JM, Vedvick TS, Reschke WC (1993) Recent advances in the expression of foreign genes in *Pichia pastoris*. Biotechnology 11:905–910

Demain AL, Gerson DF, Fang A (2005) Effective levels of tetanus toxin can be made in a production medium totally lacking both animal (e.g., brain heart infusion) and dairy proteins or digests (e.g., casein hydrolysates). Vaccine 23:5420–5423

Demain AL, George S, Kole DF, Gerson DF, Fang A (2007) Tetanus toxin production in soy-based medium: nutritional studies and scale-up into small fermentors. Lett Appl Microbiol 45:635–638

Fang A, Gerson DF, Demain AL (2006) Menstrum for culture preservation and medium for seed preparation in a tetanus toxin production process containing no animal or dairy products. Lett Appl Microbiol 43:360–363

Fields R (1971) The measurement of amino groups in proteins and peptides. Biochem J 124:581

Figueroa JA, Kimball BA, Perry KR (2008) Lagomorph and rodent responses to two protein hydrolysates. Crop Prot 27:851–854

Franchon E, Bondet V, Munier-Lehmann H, Bellalou J (2006) Multiple microfermentor battery: a versatile tool for use with automated parallel cultures of microorganisms producing recombinant proteins and for optimization of cultivation protocols. Appl Environ Microbiol 72:5225–5231

Heidemann R, Zhang C, Qi H, Rule J, Rozales C, Park S, Chuppa S, Ray M, Michaels J, Konstantinov K, Naveh D (2000) The use of peptones as medium additives for the production of a recombinant therapeutic protein in high density perfusion cultures of mammalian cells. Cytotechnology 32:157–167

Inagrosa (2002) http://www.inagrosa.es/biblioteca_i1.html. Accessed 16 June 2009

Jens Adler-Nissen (1986) Some fundamental aspects of food protein hydrolysis. In: Jens Adler-Nissen (editor) Enzymic hydrolysis of proteins, Elsevier applied science publishers, NY

Kataoka M, Hashimoto K-I, Yoshida M, Nakamatsu T, Horinouchi S, Kawasaki H (2006) Gene expression of *Corynebacterium glutamicum* in response to the conditions inducing glutamate overproduction. Lett Appl Microbiol 42:471–476

Kunert R, Gach J, Katinger H (2008) Expression of a Fab fragment in CHO and *Pichia pastoris*. A comparative case study. Bio Proc Int 6(Suppl 4):34–40

Lamsa M, Bloebaum P (1990) Mutation and screening to increase chymosin yield in a genetically-engineered strain of *Aspergillus awamori*. J Ind Microbiol 5:229–238

Naegeli C (1880) Sitz'ber, math-physik. Klasse Akad. Wiss Meunchem 10: 277

Neutra R, Levi B-Z, Shoham Y (1992) Optimization of protein production by the baculovirus expression vector system in shake flasks. Appl Microbiol Biotechnol 37:74–78

Olsen D, Chang R, Williams KE, Polarek JW (2010) The development of novel recombinant human gelatins as replacements for animal-derived gelatin hydrolysates in pharmaceutical applications. In: Pasupuleti VK, Demain AL (eds) Protein hydrolysates in biotechnology. Springer, The Netherlands

Pasupuleti VK (2006) Proteins power up. Food Technol 2:55–57

Penttila ME, Suihko M-L, Lehtinen U, Nikkola M, Knowles JKC (1987) Construction of brewer's yeasts secreting fungal endo-beta-glucanase. Curr Genet 12:413–420

Sobharani P, Agrawal R (2009 Feb 25) Supplementation of adjuvants for increasing the nutritive value and cell viability of probiotic fermented milk beverage. Int J Food Sci Nutr 1–14

Sorensen SPL (1907) Formol titration. Biochem Z 7:45

Taylor WG, Dworkin RA, Pumper RW, Evans VJ (1972) Biological efficacy of several commercially available peptones for mammalian cells in culture. Exp Cell Res 74:275–279

Taylor WG, Dworkin RA, Pumper RW, Evans VJ (1974) Studies on a serum substitute for mammalian cells in culture. In Vitro Cell Dev Biol Plant 9:278–285

Ward M, Lin C, Victoria DC, Fox BP, Fox JA et al (2004) Characterization of humanized antibodies secreted by *Aspergillus niger*. Appl Environ Microbiol 70:2567–2576

World Patent, WO 96/26266
World Patent, WO 98/15614
World Patent, WO 01/23527
World Patent, WO 00/0300
World Patent, WO 98/08934
World Patent, WO 06/045438
Zhang C-H, Wu J-Y, He G-Y (2002) Effects of inoculum size and age on biomass growth and paclitaxel production of elicitor-treated *Taxus yunnanensis* cell cultures. Appl Microbiol Biotechnol 60:396–402

Chapter 2
State of the Art Manufacturing of Protein Hydrolysates

Vijai K. Pasupuleti and Steven Braun

Abstract The use of protein hydrolysates in microbiological media has been in existence for several decades and the basic manufacturing process of protein hydrolysates has remained the same. However, with increasing use of protein hydrolysates in specialized applications such as animal cell culture processes, the manufacturing of protein hydrolysates has dramatically improved and is still in its infancy to uncover the specific peptide, peptides and combination of individual amino acids that produce intended effects for that application. This will change as the protein hydrolysate manufacturers and end-users exchange information and work towards the common goal of developing the best protein hydrolysates for specific applications. This chapter will review the generic manufacturing of protein hydrolysates describing individual unit operations, problems faced by manufacturers and suggestions for obtaining consistent product and guidelines for the end-users in getting regulatory support and setting up reliable specifications. Finally the chapter concludes with future trends of protein hydrolysates.

Keywords Manufacturing • Protein hydrolysates • Downstream processing • Inconsistencies • Hydrolysis

Introduction

The most basic function of protein hydrolysates in the applications of biotechnology is to provide a nitrogen source for bacteriological, industrial and specialized media for microbial, plant, animal and insect cell cultures on both a laboratory and industrial scale. However, in many instances protein hydrolysates also provide vitamins,

V.K. Pasupuleti (✉)
SAI International, Geneva, IL 60134, USA
e-mail: Vijai1436@sbcglobal.net

S. Braun
Mead Johnson Nutrition, Evansville, IN 47721, USA

V.K. Pasupuleti and A.L. Demain (eds.), *Protein Hydrolysates in Biotechnology*,
DOI 10.1007/978-1-4020-6674-0_2, © Springer Science+Business Media B.V. 2010

minerals and unknown growth factors resulting in higher yields and productivities (Pasupuleti and Schie 1998). For more details the reader is encouraged to refer to the Chapters 1, 3–7 for detailed discussions on use of protein hydrolysates with specific reference to industrial fermentations, cell culture and microbiological media.

The current modern manufacturing facilities with sophisticated equipment and technology for manufacturing protein hydrolysates are a result of long standing practice, extensive research and to a certain extent trial and error starting from the first attempt to grow microbial and animal cells. During the early days, the nitrogen requirements of bacteria grown in the laboratory were met by adding naturally occurring substances like blood, body fluids, etc., to the media. The first report on the use of egg albumin was published in 1882 by Naegeli, which he referred to as "peptone," a term still used today for some products. However, later it was discovered that peptones derived from the partial digestion (hydrolysis) of proteins would furnish organic nitrogen in a more available form (Peters and Snell 1953; Pasupuleti 1998). These peptones that are commonly known as protein hydrolysates are now the preferred source of nitrogen for bacteriological culture media, commercial fermentations and specialized cell culture processes. From its first use as an improved source of nitrogen in bacteriological media, protein hydrolysates are now used widely in other areas of biotechnology such as:

- Animal cell culture for the production of monoclonal antibodies, therapeutic proteins, enzymes, etc. (Pasupuleti 1998, 2001; Ganglberger et al. 2005).
- Recombinant culture fermentations for the manufacture of therapeutic drugs, vaccines, etc. (Mazurkova et al. 2008; Kwon et al. 2008; Heidemann et al. 2004; Tripathi et al. 2009).
- Insect and plant cell cultures for a variety of end products (Ikonomou et al. 2001, Kwon et al. 2005; Girón-Calle et al. 2008).
- Specialized media for growing and expressing the genetically modified micro-organisms (Blattner 1977; BD Manual 2009).
- Animal feeds for higher milk output, better meat quality, increased weight gain in shorter time frames (Backwell 1998; Choung and Chabmberlain 1998; Gilbert et al. 2008; Cordoba et al. 2005).
- Crops for enhanced yields and productivities (Kinnersley et al. 2003; El-Naggar 2006).

Many if not all of these newer applications require higher quality specialty protein hydrolysates with stringent specifications rather than the crude preparations that were initially manufactured. To meet this demand, the industry is responding with newer protein hydrolysates that are an excellent nutritional source providing di, tri, oliogpeptides and amino acids, which exert specific physiological functions to increase the productivities of cell culture and fermentations (Pasupuleti 2000). This field is rapidly growing and the sales of protein hydrolysates are increasing year after year and the manufacturers are refining their processes to better define their products. Still to a larger extent especially in the cell culture applications, much of the work is trial and error because complete characterization of protein hydrolysates with respect to the requirements of animal cells is not as well known as that seen in defined media (Pasupuleti 2007).

In this chapter, we focus mainly on the manufacturing of protein hydrolysates related to biotechnology applications and it is beyond the scope of this chapter to cover the manufacturing of protein hydrolysates for nutraceuticals and food applications. This chapter will review:

1. Current practices of different manufacturing methods of protein hydrolysates: acid, alkali, enzymatic hydrolysis of proteins and briefly about upcoming novel manufacturing of peptides by enzymes and fermentations. A generic flow of manufacturing process of protein hydrolysates will be described.
2. Selection of raw materials, protein solubilization, pretreatment methods, controlling and monitoring of microorganisms during the manufacturing, significance of proteases, the degree of hydrolysis, AN/TN ratio, peptide chain length, molecular weight distribution profiles, free and total amino acids will be highlighted.
3. Downstream processing which is dictated by the end use of protein hydrolysates: reviews brief description of plate and frame filtration and carbon treatment methods. Centrifugation, micro, ultra and nano filtration, reverse osmosis, column chromatography, ion exchange will not be covered in this chapter.
4. Different methods of concentration and drying: spray dried powders; roller drum dried powders and agglomerated powders.
5. Considerations in plant design and layout: importance of the source of water, ISO/GMP facilities.
6. Future trends.

Current Practices of Different Manufacturing Methods of Protein Hydrolysates

The predominant manufacturing method of protein hydrolysates for applications in biotechnology is by enzymes. However, protein hydrolysates made by acid and alkaline hydrolysis of proteins are also commercially available.

Acid hydrolysis of proteins: The first reported acid hydrolysis goes back to 1820 by Braconnot (Hill 1965). It took several decades for commercialization and is still in practice representing one of the older processes. The majority of acid hydrolyzed proteins are used as flavor enhancers and only a small portion of these acid hydrolysates are being used in biotechnology, sometimes after partially or completely removing the salt (Nagodawithana 1995).

Hydrochloric and sulfuric acid are mainly used to hydrolyze proteins, the most common being hydrochloric acid. With the acid hydrolysis some of the essential amino acids such as tryptophan, methionine, cystine and cysteine are destroyed. Further, glutamine and aspargine are converted to glutamic acid and aspartic acid (Bucci and Unlu 2000). Typically, acid hydrolysis breaks down the proteins into individual amino acids and minute amounts of smaller peptides. The process adds significant amounts of salt, which is detrimental to the growth of microorganisms. For this reason some of the manufacturers remove salt partially or completely by precipitation nanofiltration and/or ion exchange resins. However, acid hydrolysates are widely used in the food and pet food industry as flavor enhancers (Nagodawithana 1998, 2010). This is covered in greater detail in Chapter 11.

The manufacturing of acid hydrolysates requires glass lined stainless steel reactors that can withstand high pressure and temperatures. Equally important are the safety procedures built around these processes. The governing factors for acid hydrolysis are the concentration and type of acid (hydrochloric acid or sulfuric acid), temperature (250–280°F), pressure (32–45 psi), time of hydrolysis (2–8 h) and the concentration of protein (50–65%). All of these independently and combined will have an impact on quality of the product.

Acid hydrolysates of casein and soy proteins are commercially available from manufacturers for use in fermentations and diagnostic media.

Alkaline hydrolysis of proteins: To the best of our knowledge, there are no reported commercial applications of alkaline protein hydrolysates in biotechnology. However, in the food industry, alkaline protein hydrolysates are used on commercial scale. Some amino acids like serine and threonine are destroyed during alkaline hydrolysis but tryptophan is intact.

Alkaline hydrolysis is a fairly simple and straightforward process; first the protein is solubilized by heating followed by the addition of alkaline agents like calcium, sodium or potassium hydroxide and maintaining the temperature to a desired set point (typical range 80–130°F). The hydrolysis will be continued for several hours until it reaches the desired degree of hydrolysis and then the product is evaporated, pasteurized and spray dried.

Enzymatic hydrolysis of proteins: Most of the enzymes used to make protein hydrolysates for applications in biotechnology are obtained from animal, plant and microbial sources. Recent advances in techniques for the hydrolysis of proteins have come from studies mainly with proteolytic enzymes from fermentation processes. The main advantages of enzyme hydrolysis of proteins is that the hydrolysis conditions are mild and enzymes are more specific enabling the manufacturers to precisely control the degree of hydrolysis and tailor make products for the end users.

A wide variety of proteolytic enzymes are commercially available from animal, plant and microbial fermentations. The most commonly used enzymes for protein hydrolysates from animal sources are pancreatin, trypsin, pepsin; plant sources are papain and bromelain and from fermentation sources, bacterial and fungal proteases. The hydrolysis of proteins can be achieved by a single enzymatic step or a sequential enzyme hydrolysis using multiple enzymes. The choice of enzyme depends on the protein source and end user requirements. For example, if the protein has a higher content of hydrophobic amino acids then the enzyme of choice should be the one that preferentially cleaves the hydrophobic amino acids (Adler-Nissen 1986).

Besides acid, alkaline and enzymatic hydrolysis of proteins, a new trend in the manufacturing of protein hydrolysates for (di and tri peptides) is by coupling of the amino acids by enzymes or mass producing them by a fermentation process (Ota et al. 1999; Pasupuleti 2005).

Generic Flow of Manufacturing Process of Protein Hydrolysates

Depending on the manufacturer and type of product, the reactor size varies anywhere between 500 and >10,000 gal. As shown in the schematic flow chart in Figs. 1 and 2, first the proteins are generally solubilized in water to anywhere between 8–20% solids. If needed the proteins are pretreated with heat (up to 200°F), acid or alkali,

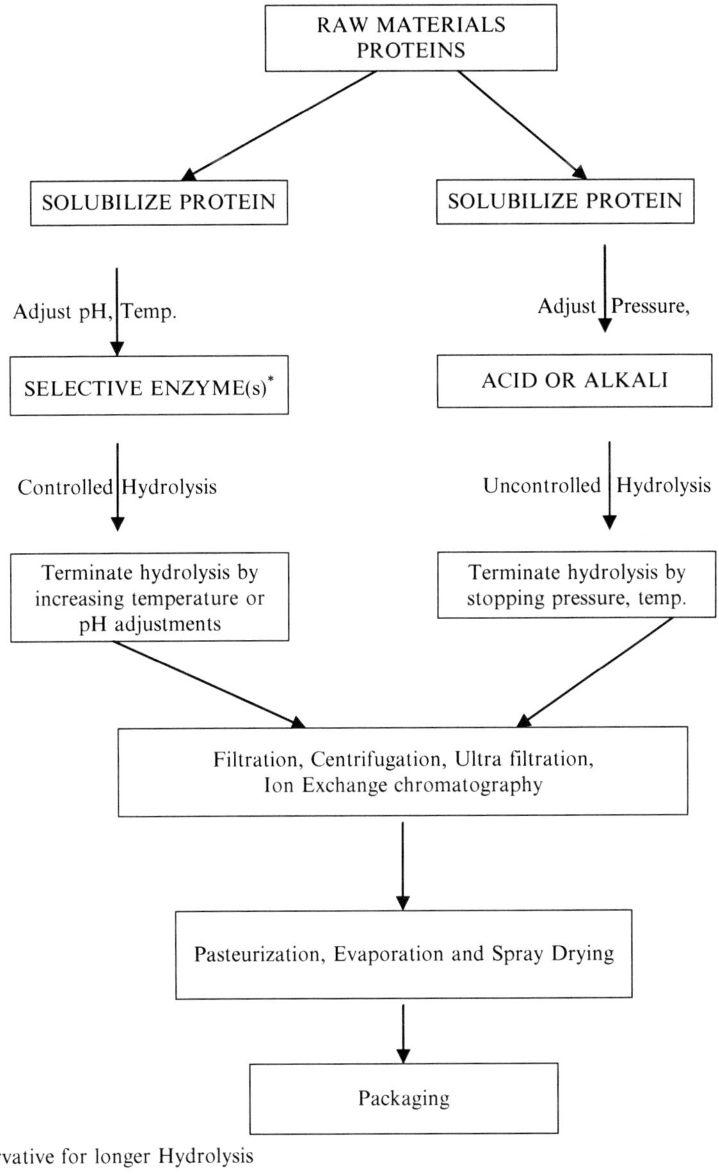

Fig. 1 Typical manufacturing overview of protein hydrolysates

Fig. 2 Bank of hydrolysis reactors (Courtesy: Kerry Biosciences)

adjusted to appropriate pH (3.5–9.0) and temperature (100–150°F) and followed by the addition of enzyme or enzyme systems. In commercial practice the enzymes used could be purified, semi-purified or crude from animal glands. The hydrolysis times vary anywhere from for 1 h to more than 100 h and some manufacturers use bacteriostatic or bactericidal preservatives especially when the hydrolysis times are longer to prevent microbial contamination. The bacteriostatic or bactericidal preservatives will be subsequently removed or evaporated during the downstream processing. Use of bacteriostatic or bactericidal preservatives is an old practice and today there are better alternatives such as UV treatment, pulse electric field applications (Uchida et al. 2008), ozone (Oliver and Duncan 2002) that need to be explored. Some protein hydrolysis manufacturers have already eliminated or are attempting to eliminate Bacteriostats in some of their processes by shortening the hydrolysis times to 4 h or using enzymes that can withstand higher temperatures or enzymes that work at acidic pH environment values. These new technologies will give better control of the process and deliver a consistent product batch after batch. The degree of hydrolysis (DH) or time in the reactor is monitored by taking in-process samples. Once the desired DH level is achieved the hydrolysis is typically terminated by heating to higher temperatures to deactivate the enzyme or enzyme systems.

Depending on end use, the protein digest is pasteurized (the term pasteurization used in this industry is different from that used in the typical dairy industry relating to time and temperature), evaporated and spray dried or more typically it goes through a series of downstream processing steps. Almost all of the products used in animal cell culture invariably go through the purification process. Typically the first step is

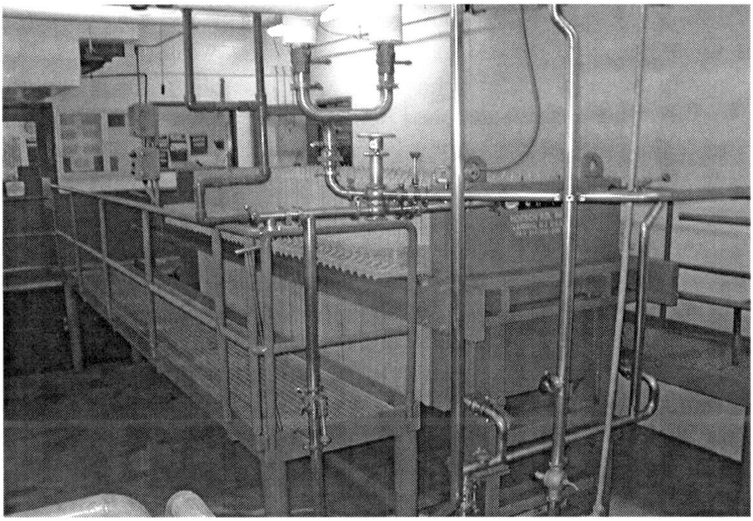

Fig. 3 Plate and frame filter press (Courtesy: Kerry Biosciences)

Fig. 4 Centrifuge (Courtesy: Kerry Biosciences)

to separate the insolubles from the protein digest by using a centrifuge or a plate and frame filter press or a micro filtration system (Figs. 3 and 4). Sometimes, the filtration process is repeated several times until a desirable color and clarity of the solution is obtained. Especially when plate and frame filter press is used, an inert filter aid (usually from volcanic rocks or diatomaceous earth) is used to form a thin coating on the plates often referred to as "precoating" for better filtration. Sometimes the filter aid is mixed with the protein digest and recycles through the filter press until a nice coat is formed before the clear filtrate is collected and this process is often referred to as "body-feed coating". Charcoal powder is commonly used to decolorize and to remove haze-forming components (Chae et al. 1998; John 1993). This unit operation is very critical to obtain desirable color and clarity. The manufacturers have an arsenal of different operating conditions such as temperature (hot or cold), pH and flow rates to achieve the desired product specifications.

The latest addition to the manufacturing of protein hydrolysates is to replace the plate and frame filter press with that of the ultra filtration systems (Fig. 5). The protein hydrolysate industry typically uses spiral wound ceramic or hollow fiber membranes with a 10,000 Da molecular weight cut off. This is to remove endotoxins as a lower amount of endotoxins is preferred for cell culture applications. However, this could also potentially remove the peptides greater than 10,000 Da. For this reason some manufacturers use 50,000 or greater molecular weight cut off membranes but this may not guarantee the removal of endotoxins. Therefore, depending on the end use, the manufacturer uses either 10,000 or higher molecular weight cut off membranes. However, in a recent publication, the authors demonstrated that endotoxin specification for cell culture may not be critical (Limke 2009).

After filtration the product is typically pasteurized or heat treated to kill/reduce the microorganisms. The term pasteurization is misleading because in most cases the temperatures used are way beyond standard legal pasteurization temperatures. Some manufacturers tend to pasteurize the product multiple times. Some manufacturers tend to pasteurize the product multiple times.

Fig. 5 Ultra filtration system (Courtesy: Milk Specialties Global)

After pasteurization the product is evaporated to remove thousands of gallons of water to bring the solids to 30–50%. Interestingly not all protein hydrolysates evaporate at the same rate and to the same level of solids. It varies from protein source, choice of enzyme and degree of hydrolysis. For this reason the bulk densities of protein hydrolysates differ.

Some manufacturers offer a concentrated protein hydrolysate, typically > 60% solids with very low water activity that is microbially stable yet dispensable through pumps. This is advantageous in terms of handling, safety and prevents the loss of protein hydrolysate as dust in batching to make the media. The other advantage is that the end user could potentially work with the manufacturers to customize the package of protein hydrolysate concentrate to their use in the fermentors. However, the end users should have the capabilities to use concentrates.

After evaporation some manufacturers go through final Pasteurization and then the product typically goes through a feed tank that feeds to the spray drier and eventually into the boxes, bags or drums for packaging.

Controlling Inconsistencies of Protein Hydrolysates During Manufacturing

The following are very important to manufacture high quality consistent protein hydrolysates:

- Strictly adhering to GMP procedures.
- Maintaining hygiene of the plant at all times.
- Screening of raw materials and qualifying vendors to obtain consistent and high quality raw materials.
- Monitoring in-process samples to maintain consistency of the batches.
- Testing finished product samples from beginning of first drum to the last drum of each batch.
- Robustness of the process dictates the consistency of protein hydrolysates.
- Constantly monitoring protein, enzyme, water sources and the downstream processing techniques as they create differences in the quality of protein hydrolysates.

As mentioned above depending on the end use, protein hydrolysates go through a variety of downstream processing steps. For example if the protein hydrolysate is specifically made for cell culture use, then typically after the hydrolysis it will go through a heating step to deactivate the enzyme followed by separation either by centrifugation or plate and frame filtration or ultra filtration, evaporation, pasteurization and spray drying. Sometimes the product fails to meet specifications especially due to color and clarity at any of these steps. When this happens the manufacturers will perform additional filtrations, adjust pH and do what is required to meet the end product specifications that are described in their standard manufacturing procedures. The goal for the manufacturer is to meet the specifications; however, by performing additional steps the quality of product may change and that may impact the performance in end users applications. Therefore, it is important to develop a well defined and robust process that will alleviate these types of problems. This can only be achieved by partnering with the end users to understand their requirements and reviewing manufacturer's capabilities and by jointly developing physical, chemical and functional specifications.

It is also important to share the process details especially if there are any variations with the end users and they in return should share the results obtained and this will help to better define protein hydrolysates.

For example a number of practises noted below are used in protein hydrolysate manufacture but these may not be acceptable for some end uses for the reasons noted.

- When the desired degree of hydrolysis is not achieved the common practice is to add more enzymes than the standard. This could potentially change the enzyme/substrate ratio and subsequently the quality of the product.

  Sometimes this happens and the reason could be that the enzyme used from animal glands is not active. If this is the case move towards purified enzymes that have consistent activity.

- The hydrolysis times are not consistent suggesting inconsistencies and therefore it is not a robust process.

  If the enzyme activity is not consistent, add enzymes based on their activity and not by quantity.

- Goes through pH swings because of operator/machine errors.

  Validate the machines periodically and train operators.

- Performs multiple filtrations using inert filter aid and/or charcoal powder.

  Avoid this as it may potentially remove some of the amino acids and peptides. If this has to be performed make sure to check the functionality of end users' applications.

- To obtain the desired color and clarity sometimes more charcoal is added.

  Avoid this as it may potentially remove some of the amino acids and peptides. If this has to be performed make sure to check the functionality of end users' applications.

- Mixing remains of old batches of the product into the new batch during processing.

  Avoid this as it may potentially remove some of the amino acids and peptides. If this has to be performed make sure to check the functionality of end users' applications.

- Mix rejected product into the new batch in hopes to meet the desired specifications.

  Avoid this as it may potentially remove some of the amino acids and peptides. If this has to be performed make sure to check the functionality of end users' applications.

- Dry powder blend with other batches. For example, if one batch fails to meet the desired AN/TN specification of 45 and reaches only to 40. This will be potentially mixed with the next batch that is tailor made to be blended with high AN/TN 50 or greater to get to the specification of 45.

  Avoid this as it may potentially remove some of the amino acids and peptides. If this has to be performed make sure to check the functionality of end users' applications.

- The entire batch that was previously rejected for not meeting the specifications will be reworked into a new batch and sold as a new batch.

  Avoid this as it may potentially remove some of the amino acids and peptides. If this has to be performed make sure to check the functionality of end users' applications.

- Because of unanticipated problems in the processing and/or scheduling conflicts sometimes the product is held for several hours. If the hold times are unusually longer then preservatives are added which could potentially change the quality of the product.

  Avoid this as it may potentially remove some of the amino acids and peptides. If this has to be performed make sure to check the functionality of end users' applications.

All or some of the above practices may be acceptable in certain applications where the product functionality may not be very specific to protein hydrolysates. Indeed if any or all of the above is used for cell culture applications it is important that the end user understand the process used and understand the impact on the cell culture performance. Sharing this information with end-users may lead to a better understanding of protein hydrolysates and the cell culture performance. This will ultimately benefit the end user as well as the manufacturer.

Selection of Raw Materials

The most commonly used animal protein hydrolysates in biotechnology applications are casein, whey and meat obtained from different organs. Widely used plant derived proteins are from soy and wheat however; recently rice, pea and cottonseed proteins have been introduced commercially (Sheffield Product Manual 2009). It is important to note that the amino acid composition, structure of the protein, its solubility and denaturation will affect the choice of enzyme and the subsequent product.

High quality and consistent protein hydrolysates start with the selection of right raw materials. It is for this reason the manufacturers should spend a great deal of time and resources to ensure their availability, note seasonal fluctuations if any, meet regulatory requirements (BSE/TSE free, Kosher, Halal, etc.) and qualify multiple vendors.

Besides protein source, the most important raw materials are enzymes, water source and process aids as they have biggest impact on the end product. Therefore, it is essential to thoroughly study these sources, establish rigorous quality control specifications, audit and qualify a minimum of two vendors.

Enzymes: The degree of hydrolysis (AN/TN ratio), peptide chain length, molecular weight distribution profiles, free and total amino acids are dependent on the choice of enzyme or enzyme systems.

Compared to acid and alkaline hydrolysis of proteins, proteolytic enzymes offer significant advantages such as:

- Requirement of small amounts of enzymes that can be easily deactivated after the hydrolysis.
- Mild operating conditions like temperature and pH.
- No destruction of amino acids.
- No production of chloropropanols.
- Hydrolysis of certain amino acids resulting in less complex mixtures of peptides that can be relatively easily purified.
- Availability of several choices of enzymes. This enables the manufacturer to pick the best one to specifically modify the proteins that best suits end-user applications.

The use of proteolytic enzymes or proteases that specifically break down the proteins into peptides with different peptide chain lengths and free amino acids has been in practice for more than several decades. In the beginning, most of the proteases used to hydrolyze proteins for use in the growth of microorganisms were from animal sources especially porcine. Interestingly enough, even today these enzymes are widely used individually or in mixtures. Some of the common examples are pepsin, pancreatin, carboxy peptidases and amino peptidases. At the same time, proteases from plant sources are also widely used, for example, papain and bromelain. Fermentation technology and genetic engineering disciplines have advanced over the years and yielded novel enzymes. Few examples, serine proteases, fungal proteases, endo and exo peptidases.

The industrial practice of defining protein hydrolysates is by determining the degree of hydrolysis of proteins which is the percentage of peptide bonds cleaved form a given protein; refer to Chapter 1 for the definition of degree of hydrolysis. The protein hydrolysates used in biotechnology are often described by the ratio of Amino Nitrogen to Total Nitrogen (AN/TN). The most commonly used methods to determine AN are by:

1. Formal Titration (Sorensen 1908) Opa (H. Frister, H. Meisel, E. Schlimme)
2. OPA method modified by use of N,N-dimethyl2-mercaptoehtyl ammonium chloride as thiol component (Fres 1988)
3. TNBS (Adler-Nissen 1979)

The total nitrogen is determined by traditional Kjeldahl (Bradstreet 1954) or modified Kjeldahl (Fearon 1920) or combustion methods (Brink and Sebranek 1993).

Further, protein hydrolysates defined by determining the amount of free and total amino acids; molecular weight distribution of peptides (Jandik et al. 2003; Jun et al. 2008).

End Use of Protein Hydrolysates Dictates the Downstream Processing

Choosing the right protein, enzyme(s) and obtaining the desired degree of hydrolysis is only half the battle. Equally important is the downstream processing as it dictates the end use of protein hydrolysates. For example, after protein hydrolysis, it goes through one simple separation step and the resulting product may not be clear and soluble. This product will not meet the demands of dehydrated culture media or cell culture applications but may be good enough as a nitrogen source for some industrial fermentations. Similarly a product that did not go through ultra filtration with the 10,000 molecular weight cut off and was neither monitored nor controlled for endotoxins may not be good for certain cell culture applications.

Some of the commonly used equipment for purification are:

- Centrifugation
- Plate and Frame Filtration
- Micro Filtration
- Ultra Filtration
- Nano Filtration
- Ion Exchange Chromatography

The detailed description of most of the above equipment is beyond the scope of this chapter and can be found in the literature as they are commonly used in biotechnology. However, a brief description of plate and frame filtration will be given to make the reader familiar with it, how and why it is still commonly employed by some of the protein hydrolysate manufacturers.

Plate and Frame Filtration

Essentially plate and frame filter presses are dewatering machines that work under pressure typically at 60–80 psi to remove the insolubles from protein hydrolysate digests. They are particularly suited in cases where insoluble solids are higher that may potentially plug the micro and ultra filtration membrane systems.

Typically diatomaceous earth is used with plate and frame filters to aid in the filtration of very fine solids. This coating is first applied to the filter, allowing an even coating of the filter cloth; then the feed is introduced into the press or sometimes the diatomaceous earth is mixed with the protein hydrolysates slurry and run through the filter until a fine coating is formed. During this time the filtrate is recycled back to the protein hydrolysate digest. This will typically take about 15–30 min and once the coating is formed, the filtrate will be collected. The biggest advantage of using the plate and frame filter is that it allows removal of color during filtration by treating with various grades of carbon. However, it is important to pay attention to the type and quantity of carbon applied as it may potentially remove some of the peptides and amino acids (Silvestre et al. 2009).

Figure 3 shows the basic operation of a plate and frame filter press. The feed enters the filter press at the bottom of the plate, using a pump that is capable of creating 80–90 psi. Then, the feed travels the path of least resistance (up between the filter plates, which has filter media inserted between the plates), and the void between the plates is filled with the slurry, as the liquid passes through the filter media, and travels up to the outlet port at the top of the plate. This liquid is referred to as the "filtrate", and is discharged from the filter press. The solids remain in the void between the plates, until the plates discharge the filtered solids (Mine Engineer Product Manual 2009).

Large plate and frame filter presses have mechanical "plate shifters", to move the plates, allowing the rapid discharge of the solids stuck in between them. Also, they have the capability of blowing compressed air into the plate, to dry the cake, and collect the filtrate as much as possible. Typical capacities for a plate and frame filter will depend upon the solids being dewatered. However, they will range around 1/2 gal per min of feed for low solids content slurries (<1% solids) to over 1 gal per min of slurry per square foot of surface area on the plates. For example, a 50 ft^2 filter press would dewater 25 GPM of 1/2% solids feed, or about 50 gpm of 10% solids feed (Mine Engineer Product Manual 2009). The volume between the plates will dictate how often the filter press should be stopped in order remove solids from the plates. Solids removed from the filter press typically fall into a hopper or directly onto a conveyor belt for further transport to the next stage of the operation. The filtrate goes through a series of downstream processing steps depending on the end use application. A 10 ft^3 volume press would hold 10 ft^3 of dewatered solids.

Many protein hydrolysate manufacturers first use a laboratory or small scale pilot plant plate and frame filter press to determine and a optimize the filtration in terms of color, clarity, throughput and yields. The smaller versions are exactly like a full scale press, and results obtained can be scaled up for plant operations, accurately.

Depending on the end use and manufacturers capabilities, sometimes after the first plate and frame filtration, the filtrate is passed through micro and ultra filtration systems. It is also a common practice of some manufacturers to use plate and frame as the last filtration step to obtain the desired color and clarity. In cases where very low salts are desired, the filtrate may go through ion exchange chromatography to remove the salts.

Different Methods of Concentration and Drying

Several methods are employed to concentrate the protein hydrolysates such as falling film, rising film evaporators. In some instances nano filtration is also used to concentrate (Fig. 6).

Depending on the nature of protein hydrolysates, they are concentrated from 25–>50% solids before drying. In some instances, the manufacturers may not dry

Fig. 6 Falling film evaporator (Courtesy: Niro Inc.)

and offer protein hydrolysates in a concentrate form that is pumpable at > 60% solids. The water activities of these concentrates are kept low so that they are microbially stable.

Once the product is concentrated, it is pumped to a feed tank that goes through a cartridge filter to remove any larger particles, then to a high pressure pump to feed the dryer. Spray driers are widely used in the industry and some manufacturers use roller drum driers (Figs. 7 and 8).

Considerations for Plant Design and Layout

Most of the protein hydrolysates manufacturers have been in the business for several decades and so are their plants. They are very good in adding new equipment to stay up with the latest technology and to increase productivities.

Whether it is an existing or a new plant, the most important things to consider are safety, hygiene, water source. If needed treat the water with appropriate technologies such as reverse osmosis or water softeners. Use proper drainage systems, handling of solid and liquid effluents and right size tanks with appropriate design for mixing. If direct steam is used to heat the tanks, they should be able to withstand steam injection. It is important to use downstream processing equipment to match the size of the tanks, strictly complying with GMP procedures.

Since protein hydrolysates are made from a wide variety of proteins including meat and animal tissues, it is important to determine if the vendors can provide documentation and answer customers' questions on regulatory issues. If yes, to what extent is the support available; and the certificate of suitability from the European Commission.

Fig. 7 Spray dryer (Courtesy: Niro Inc.)

Concerns of the Biotechnology Industry for Regulatory Support from Protein Hydrolysate Manufacturers

- Is there adequate separation of animal origin products from that of the vegetable proteins?
- Are the raw materials and finished products appropriately labeled and stored to prevent mix-ups?
- How are the raw materials and finished products that did not pass the QC tests segregated?
- Are discrepancy reports issued for the products that did not meet the required specifications and written procedures describing how this product is disposed of?
- Is there a quarantine location to hold the products before QA/QC releases the raw materials and finished product?
- Are there written acceptance specifications for raw materials and finished products?

Fig. 8 Drum dryer (Courtesy: Nutraflo Protein Products)

- How the cross contamination between animal origin and vegetable proteins is prevented?
- Are there good written cleaning procedures in place?
- Are the cleaning procedures validated and documented?
- What measures are taken to ensure that the cleaning agent is completely removed from the process tanks, lines and equipment?
- If preservatives are used in the manufacturing, what procedures are followed to ensure that they are totally removed from the product?
- What tests are performed to ensure that the enzyme is totally deactivated?
- Does QA/QC report to somebody other than manufacturing?
- Is the process reproducible and validated?
- Is the process equipment calibrated and validated?

- Are there written Manufacturing Standard Operating Procedures?
- Is the water periodically tested and are there written specifications?
- Is the manufacturing plant designed to control bioburdens, endotoxins and mix-ups with other products?
- What is the lot size and is it produced by following the approved written production process and formulations?
- Are the test procedures validated and documented?
- Are instruments calibrated by following a written schedule and against standards traceable to the NIST?
- Is the process equipment sufficiently cleaned, tested and documented so that the product from previous runs is not getting mixed up?
- Is the temperature and humidity of the warehouse monitored constantly and documented?
- Change control procedures – who is responsible?

Specifications and Sampling

It is very important that the specifications and sampling should be relevant to the end users and should match the capabilities of the manufacturers.

Typically, protein hydrolysate manufacturers will take in-process and finished product samples to ensure that the product is in compliance with the established specifications for that particular product.

It is important that these specifications are relevant to your applications. The most important aspect of sampling is how representative the sample is of the entire lot. The lot varies anywhere from 1,000 to >10,000 pounds. There is a possibility that the product from container #1 could be different from the last container because of the inherent long processing times. Therefore, it is reasonable to ask the manufacturer about sampling plans and ensure that a representative sample of the entire lot is taken. At a minimum, ask manufacturers to take samples from the first, middle and the last drum and blend them together before use.

If the sample is not uniform with the whole lot, then the chances are that the customer may get poor results with their trials and may not trust the data.

What specifications and test methods should be established for cell culture?

It is critical to minimize the variations and deliver consistent protein hydrolysate batch after batch.

For recommended tests, it is important that both the laboratories should follow the same test method to avoid confusion. Important test methods are:

- Solubility; define appropriately and perform the test at 50°C.
- Clarity with 5% solids using Hach turbidometer.
- Color with 2% solids at 420 nm.
- Filterability with Vmax filterability test method.
- Endotoxins with Gel clot method and reported as EU/g (Tsuji et al 1980).

- AN with Sorensen or Formaldehyde titration method or OPA or TNBS method.
- TN with Kjeldahl or direct combustion method.
- Minerals with atomic absorptiometer.

These are only a few guidelines and the end users have to customize the specifications and test methods to suit their needs. Clearly communicate and get agreement with the protein hydrolysate manufacturers.

Manufacturing capabilities should go beyond meeting the product specifications. As in some applications, for example in cell culture, protein hydrolysates significantly enhance the yields but its mechanism is not clearly established. In such cases, it is essential that the manufacturing of protein hydrolysates be highly consistent with respect to the process starting from raw materials, pretreatment, hydrolysis times, filtration, pasteurization, evaporation, spray drying times to the packaging. Clear understanding of manufacturing capabilities in meeting the product specifications and understanding end users' applications is important. Sharing knowledge and collaborating with each other will lead to better products and increased productivities.

Future Trends

Continuous Process

To the best of our knowledge, all the protein hydrolysate manufacturers employ traditional batch systems. However, there is a potential to use immobilized enzyme or membrane bioreactor systems to save the enzyme cost and increase the productivities (Holownia 2008).

Innovative Technologies

Manufacturing of protein hydrolysates, peptides by fermentation process. An example is glutamine dipeptide. Potentially this technology could be used to make a variety of dipeptides and oligopeptides (Yagasaki 2009). The other example is manufacturing of gelatin hydrolysates by a fermentation process in response to the BSE problem (see Chapter 12 by Olsen et al. 2010).

Another revolutionary technology is the enzyme coupling method to manufacture peptides. The amino acid is esterified, and the ester is enzymatically coupled with another amino acid to make a dipeptide and so on. Conventional production methods require complicated production processes that often produce a high level of impurities (Ajinomoto Product 2009).

We have already seen a shift in using more and more enzymes obtained from fermentation process. This ensures that the enzyme activity is well defined. Enzyme manufacturers can tailor make enzymes to deliver specific protein hydrolysates for a specific function.

New equipment such as electro dialysis, nano filtration may have potential use for manufacturing protein hydrolysates (Orue 1998; Laurant and Loubna 2009).

Non-dusting protein hydrolysates may play a key role in the future as it eliminates or minimizes dust.

Collaboration and Partnerships

One of the essential elements in successful manufacturing of protein hydrolysates for biotechnological applications is working as a team in partnership with the manufacturers as well as the end users.

This type of partnership may result in new and innovative products. For example, there is a distinct possibility that protein hydrolysate manufacturers may develop one single product that may help the end users eliminate maintaining a "laundry list" of several media ingredients. They may even customize the protein hydrolysates along with other media ingredients to fit the end user fermentor size.

References

Adler-Nissen JA (1979) Determination of the degree of hydrolysis of food protein hydrolysates by trinitro benzene sulfonic acid. J Agric Food Chem 27:1256–1262

Adler-Nissen JA (ed) (1986) Enzymatic hydrolysis of food proteins. Elsevier, London

Ajinomoto Product Brochure (2009) Web site last accessed 21 Dec 2009. http://www.bioresearchonline.com/product.mvc/L-Alanyl-L-Glutamine-Amino-Acid-Ala-Gln-0001

Backwell CFR (1998) Circulating peptides and their role in milk protein síntesis. In: Grimble GK, Backwell FRC (eds) Peptides in mammalian protein metabolism. Portland Press, London, pp 69–78

BD Bionutrients Technical Manual (2009) Web site last accessed 21 Dec 2009. http://www.bdbiosciences.com/documents/bionutrients_tech_manual.pdf

Blattner FR (1977) Charon phages: safer derivatives of bacteriophage lambda for DNA cloning. Science 196:161–169

Bradstreet RB (1954) Kjeldahl method for organic nitrogen. Anal Chem 26:185–187

Brink KM, Sebranek JG (1993) Combustion method for determination of crude protein in meat and meat products: collaborative study. J AOAC Int 76:787–793

Bucci LR, Unlu L (2000) Protein and amino acid supplements in exercise and sport. In: Wolinsky I, Driskell JA (eds) Energy-yielding macronutrients and energy metabolism in sports nutrition. CRC Press, Boca Raton, FL, pp 191–212

Chae JH, In JM, Kim HM (1998) Process development for the enzymatic hydrolysis of food protein: effects of pre-treatment and post-treatments on degree of hydrolysis and other product characteristics. Biotechnol Bioprocess Eng 3:35–39

Choung JJ, Chabmberlain DG (1998) Circulating peptides and their role in milk protein síntesis. In: Grimble GK, Backwell FRC (eds) Peptides in mammalian protein metabolism. Portland Press, London, pp 79–90

Cordoba X, Borda E, Puig M (2005) Soy oligopeptides in weaning nutrition. Feed Int 26:14–18
El-Naggar AH (2006) Response of plants to natural protein hydrolysates as a nitrógeno fertilizar and chelating agent in organic agricultural systems. MS Thesis, The Royal Veterinary and Agricultural University, Denmark
Fearon WA (1920) A modified kjeldahl method for the estimation of nitrogen. Dublin J Med Sci 1:28–32
Fres Z (1988) OPA method modified by use of N, N-dimethyl 2-mercaptoehtyl ammonium chloride as thiol component. Anal Chem 330:631–633
Ganglberger P, Obermüller B, Kainer M, Hinterleitner P, Doblhoff O, Landauer K (2005) Optimization of culture medium with the use of protein hydrolysates. Cell technology for cell products. Proceedings of the 19th ESACT meeting, Harrogate, UK
Gilbert ER, Wong EA, Wrbb KE (2008) Peptide absorption and utilization: implications for animal nutrition and health. J Anim Sci 86:2135–2155
Girón-Calle J, Vioque J, Pedroche J, Alaiz M, Yust MM, Megías C, Millán F (2008) Chickpea protein hydrolysate as a substitute for serum in cell culture. Cytotechnology 57:263–272
Heidemann R, Zhang C, Qi H, Rule JL, Rozales C, Park S, Chuppa S, Ray M, Michaels J, Konstantinov K, Naveh D (2004) The use of peptones as medium additives for the production of a recombinant therapeutic protein in high density perfusion cultures of mammalian cells. Cytotechnology 32:157–167
Hill RL (1965) Hydrolysis of proteins. Adv Protein Chem 20:37–107
Holownia T (2008) Production of protein hydrolysates in an enzymatic membrane reactor. Biochem Eng J 39:221–229
Ikonomou L, Bastin G, Schneider Y-J, Agathos SN (2001) Design of an efficient medium for insect cell growth and recombinant protein production. In Vitro Cell Dev Biol Anim 37:549–559
Jandik P, Cheng J, Avalovic N (2003) Amino acid analysis in protein hydrolysates using anion exchange chromatography and IPAD. Method Mol Biol 211:155–167
John TG (1993) US Patent 5266685 - Non-bitter protein hydrolyzates
Jun LL, Chuhan-He Z, Zheng Z (2008) Analyzing molecular weight distribution of whey protein hydrolysates. Food Bioprod Process 86:1–6
Kinnersley AM, Bauer BA, Crabtree KL, Kinnersley C-Y, McIntyre JL, Daniels SE (2003) US Patent 6534446 Method to mitigate plant stress
Kwon MS, Dojima T, Park EY (2005) Use of plant-derived protein hydrolysates for enhancing growth of *Bombyx mori* (silkworm) insect cells in suspension culture. Biotechnol Appl Biochem 41:1–7
Kwon YL, Yeul SK, Heon KK, Chun BK, Lee KH, Oh DJ, Chung N (2008) Use of soybean protein hydrolysates for promoting proliferation of human keratinocytes in serum-free medium. Biotechnol Letts 30:1931–1936
Laurant B, Loubna F (2009) Membrane processes and devices for separation of bioactive peptides. Recent Pat Biotechnol 3:61–72
Limke T (2009) Impact of ultrafiltration of hydrolysates. Gen Eng News 29:29–30
Mazurkova NA, Kolokol'tsova TD, Nechaeva EA, Shishkina LN, Sergeev AN (2008) The use of components of plant origin in the development of production technology for live cold-adapted cultural influenza vaccine. Bull Exp Biol Med 146:144–147
Mine Engineer Product Manual (2009) Web site last accessed 21 Dec 2009. http://www.mine-engineer.com/mining/plate.htm
Naegeli C (1882) Cited from Hucker and Carpenter. J Inf Dis 40: 485–496, 1927. Ernhrung der neidern Pilze durch Kahlenstoff und Stickstoffuerbindenger Untermuchungen Uber Niedere. Pilze 1
Nagodawithana T (ed) (1995) Savory flavors. Esteekay, Milwaukee, WI
Nagodawithana TW (1998) Production of flavors. In: Nagodawithana TW, Reed G (eds) Nutritional requirements of commercially important microorganisms. Esteekay Associates, Milwaukee, WI, pp 298–325
Nagodawithana TW, Nelles L, Trivedi NB (2010) Protein hydrolysates as hypoallergenic, flavors and palatants for companion animals. In: Pasupuleti VK, Demain A (eds) Protein hydrolysates in biotechnology. Springer, The Netherlands

Oliver JM, Duncan HG (2002) US Patent 6387241 Method of sterilization using ozone

Olsen D, Chang R, Willimas KE, Polarek JW (2010) The development of novel recombinant human gelatins as replacements for animal-derived gelatin in pharmaceutical applications. In: Pasupuleti VK, Demain A (eds) Protein hydrolysates in biotechnology. Springer, The Netherlands

Orue MC, Bouhallab S, Garem A (1998) Nanofiltration of amino acid and peptide solutions: mechanisms of separation. J Membrane Sci 142:225–233

Ota M, Sawa A, Nio N, Ariyoshi Y (1999) Enzymatic ligation for synthesis of single-chain analogue of monellin by transglutaminase. Biopolymers 50:193–200

Pasupuleti VK (1998) Applications of protein hydrolysates in industrial fermentations. Presented at industrial and fermentation microbiology symposium, LaCrosse, WI

Pasupuleti VK (2000) Influence of protein hydrolysates on the growth of hybridomas and the production of monoclonal antibodies. Presented at the waterside conference, Miami, FL

Pasupuleti VK (2001) Commercial report on protein hydrolysates and monoclonal antibodies. SAI International, Geneva, IL

Pasupuleti VK (2005) Manufacturing of protein hydrolysates and bioactive peptides. Presented at the annual meeting of Institute of Food Technologists, New Orleans, LA

Pasupuleti VK (2007) Overview of manufacturing, characterization and screening of protein hydrolysates for industrial media formulations. Presented at Society of Industrial Microorganisms annual meeting, Denver, CO

Pasupuleti VK, Schie BJ (1998) Production of enzymes. In: Nagodawithana TW, Reed G (eds) Nutritional requirements of commercially important microorganisms. Esteekay, Milwaukee, WI, pp 129–162

Peters JV, Snell EE (1953) Peptides and bacterial growth. J Biol Chem 67:69–76

Sheffield Product Manual (2009) Web site last accessed 21 Dec 2009. http://www.sheffield-products.com/index

Silvestre MPC, Vieira CR, Silva MR, Carreira RL, Silva VDM, Morais HA (2009) Protein extraction and preparation of protein hydrolysates from rice with low phenylalanine content. Asian J Sci Res 2:146–154

Sorensen SPL (1908) Enzymestudien, Uber die quantitative Messung Proteolytischer Spaltungen. Die Formol Titrierung. Biochem Z 7:45–101

Tripathi NK, Shrivastva A, Biswal CK, Lakshmana Rao PV (2009) Optimization of culture medium for production of recombinant dengue protein in *Escherichia coli*. Ind Biotechnol 5(3):179–183

Tsuji K, Steindler KA, Harrison SJ (1980) Limulus amoebocyte lysate assay for detection and quantitation of endotoxin in a small-volume parenteral product. Appl Environ Microbiol 40:533–538

Uchida S, Houjo M, Tochikubo M (2008) Efficient sterilization of bacteria by pulse electric field in micro-gap. J Electrostat 66:427–431

Yagasaki M (2009) Fermentation technology breakthrough for the formation of dipeptides. Presented at Nutracon 2009, Anaheim, CA

Chapter 3
Towards an Understanding of How Protein Hydrolysates Stimulate More Efficient Biosynthesis in Cultured Cells

André Siemensma, James Babcock, Chris Wilcox, and Hans Huttinga

Abstract In the light of the growing demand for high quality plant-derived hydrolysates (i.e., HyPep™ and UltraPep™ series), Sheffield Bio-Science has developed a new hydrolysate platform that addresses the need for animal-free cell culture medium supplements while also minimizing variability concerns. The platform is based upon a novel approach to enzymatic digestion and more refined processing. At the heart of the platform is a rationally designed animal component-free (ACF) enzyme cocktail that includes both proteases and non-proteolytic enzymes (hydrolases) whose activities can also liberate primary components of the polymerized non-protein portion of the raw material. This enzyme system is added during a highly optimized process step that targets specific enzyme-substrate reactions to expand the range of beneficial nutritional factors made available to cells in culture. Such factors are fundamental to improving the bio-performance of the culture system, as they provide not merely growth-promoting peptides and amino acids, but also key carbohydrates, lipids, minerals, and vitamins that improve both rate and quality of protein expression, and serve to improve culture life due to osmoprotectant and anti-apoptotic properties. Also of significant note is that, compared to typical hydrolysates, the production process is greatly reduced and requires fewer steps, intrinsically yielding a better-controlled and therefore more reproducible product. Finally, the more sophisticated approach to enzymatic digestion renders hydrolysates more amenable to sterile filtration, allowing hydrolysate end users to experience streamlined media preparation and bioreactor supplementation activities. Current and future development activities will evolve from a better understanding

A. Siemensma and H. Huttinga (✉)
Sheffield Bio-Science, Veluwezoom 62, 1327 AH Almere, The Netherlands
e-mail: hans.huttinga@kerry.com

J. Babcock
Sheffield Bio-Science, 283 Langmuir Lab, 95 Brown Road, Ithaca, NY 14850, USA

C. Wilcox
Sheffield Bio-Science, 3400 Millington Road, Beloit, WI 53511, USA

V.K. Pasupuleti and A.L. Demain (eds.), *Protein Hydrolysates in Biotechnology*,
DOI 10.1007/978-1-4020-6674-0_3,

of the complex interactions within a handful of key biochemical pathways that impact the growth and productivity of industrially relevant organisms. Presented in this chapter are some examples of the efforts that have been made so far to elucidate the mechanisms for the often dramatic benefits that hydrolysates can impart on cell culture processes. Given the variety of roles that hydrolysates likely play in each cell type, close collaboration between protein hydrolysate manufacturers and biopharmaceutical developers will continue to be critical to expanding the industry's knowledge and retaining hydrolysates as a tool for enhancing media formulations.

Keywords Animal cell culture • Serum • Serum free • Monoclonal antibodies • Protein hydrolysates

Introduction

The application of recombinant technologies in mammalian cell cultures has opened the door for a multitude of new bio therapeutics including vaccines, antibodies, interferons, clotting factors, etc. Mammalian cells are the preferred production organisms to produce recombinant proteins as they can correctly fold and post-translationally modify proteins in their native, fully glycosylated form that simpler organisms cannot do. However, compared to yeasts and prokaryotes, the overall cellular production of recombinant protein is low. Therefore, the main target in bioprocess development for mammalian cell cultures is the enhancement of product yield to achieve economical and competitive concentrations of the active product. For the production of biopharmaceutical proteins with animal cells, media are often used that contain components of animal origin including bovine serum. Since these components may contain infectious agents, industry aims at using media that are animal free.

The ideal cell culture medium should allow mammalian cells to grow at the same rate as with the inclusion of sera without any loss in product yield. Accordingly, many serum-free alternatives have been developed since the late 1970s. The first step toward serum-free media (SFM) was the use of animal-derived protein hydrolysates, produced with animal-derived enzymes and/or purified proteins from animal or human sources. The search for a growth medium that most closely replicates the benefits of animal-derived sera without its attendant bio safety risks provoked a growing interest in the use of plant-based protein hydrolysates such as those derived from soy, wheat, rice, pea and cottonseed. It also appears that most actual serum and protein-free media are designed for a specific cell line are unable to sustain the cultivation of wide ranges of animal cell lines without the addition of various proteins or protein hydrolysates ('peptones').

Protein hydrolysates are now routinely employed in mammalian cell systems to enhance the overall performance of many biopharmaceutical production systems and the beneficial effect of protein hydrolysates on the growth of animal cell cultures has been known for more than 3 decades (Mizrahi 1977). Furthermore, the relative low cost of protein hydrolysates makes them attractive for application in commercial processes. The large scale cultivation of animal cells for the production of biopharmaceuticals benefits from serum replacement withdrawal in terms of production costs, process reproducibility, and ease of downstream processing, as well as bio safety. However, improving product quantity and quality requires an understanding of how protein hydrolysates (peptones) influence cell growth, maintenance, and energy metabolism. Protein hydrolysates are known to contain many components already found in standard basal media including, but not limited to, peptides, free amino acids, and carbohydrates. The manifestation of this enhancement is subject to the additive effect of the protein hydrolysate components on the final composition of the supplemented basal medium. Consequently, it is necessary to experimentally determine the proper hydrolysate dosage for a given hydrolysate-medium combination which provides the desired optimization effect, be it better growth promotion, enhanced cell viability or increased target protein production, or a combination of all the three, as determined by the requirements for a particular production system.

History of Cell Culture Media

Early biologists in the 1830s generally attempted to grow microorganisms using the food or sample on which the organism had first been observed. Later on in the 1880s, Robert Koch found that broths based on fresh beef serum or meat extracts (bouillons) gave the best growth (Koch 1876). This was further improved by the addition of a peptone (an enzymatic digest of meat) and salt to Koch's basic meat extract formulation (Loeffler 1884). By the 1890s, the culture media were developed as are known today, with Petri dishes, peptones and agar. Hydrolyzing proteins with either acid or enzymes can produce peptones. The earliest reported attempt to develop a chemically defined culture medium was done by White in 1946 (White 1946). The medium contained a mixture of amino acids, vitamins, a carbohydrate, inorganic ions, and hormones such as insulin and thyroxin. In the 1960s, Ham (1962) and his group investigated the nutritional requirements of various cells in detail and established many chemically defined media that consisted of 50–70 ingredients to eliminate or reduce the amount of serum required. In the 1970s, Sato (Hayashi and Sato 1976) and co-workers developed the hypothesis that the function of the serum was to supply hormones and growth factors required by the cells. Due to the recent progress in biotechnology, the need to cultivate mammalian cells on a large scale was increased to produce medically important substances. The basal mammalian cell medium consists of amino acids, vitamins, inorganic salts, carbon sources, and other compounds. Many basal media have been established for serum-supplemented cultures whereas some others were developed for serum-free purposes. Since many of these were developed for serum-supplemented cultures, they sometimes have to be modified for serum-free cultures. Eagle's MEM and Dulbecco's modified Eagle's medium (DMEM) are often used as the base to establish serum-free media (Dulbecco and Freeman 1959). Iscove's modified Dulbecco's medium (IMDM) was developed for the serum-free culture of lymphoid cells and has been used widely (Iscove and Melchers 1978). RPMI 1640 is also used quite often for lymphoid cell lines (Carney et al. 1981) and Ham's F12 medium is favored for epithelial cells (Ham 1962).

Although the development of serum-free media was started already in the 1970s, these media became of greater interest about 10–15 years ago. Since most basal media have been developed to be used in combination with serum, corrections of the ingredients are sometimes necessary for serum-free cultures. The first step in the development of serum-free media was the replacement of serum by protein hydrolysates of bovine origin such as Primatone™ RL and the addition of several animal-derived proteins such as the so-called ITES media (insulin, transferrin, ethanolamine, and selenium). This addition was developed for the growth of human lymphoblastoid and hybridoma cell lines and for the production of monoclonal antibodies (Murakami et al. 1982). Such media make use of various basal synthetic media such as MEM, RPMI 1640, DMEM, DMEM/F12, IMDM/F12, RPMI 1640/DMEM/F12, IMDM/F12/NCTC 135, etc. and which are supplemented with many different proteins and non-proteinous additives, are still in use today. However, the tendency has been directed toward protein and animal-free media. Suppression of the use of bovine serum is nowadays enforced because of the well-reported drawbacks of its use (Werner 2004), which precludes the approval of its use for the production of therapeutics (FDA 1997; Derouazi et al. 2004).

There is no such thing as a universal medium for optimal growth for all type of cells in culture. Culture medium optimization will always involve some degree of trial and error. Media optimization positively affects media costs and potentially creates higher-quality products in a more robust process, in addition to the ability to run smaller or infrequent batches. Because of its complexity and high cost, optimization of media formulation is a key aspect for bioprocess development in animal cell cultivation. Plant-derived protein hydrolysates (cottonseed, soy, wheat) used individually or in combination are often suitable alternatives for bovine serum to complete optimization of a lot of cell culture systems. In systems where replacement of animal-origin materials is especially difficult, using a porcine-derived peptone such as Primatone P37 may be an acceptable and efficient alternative to bovine-derived components.

Animal Cell Culture Developments

The development of cell culture was needed to produce antiviral vaccines and to understand the growth of tumor cells. Although animal cell cultures have been important at a laboratory scale for most of the last 100 years, it was the initial need for human viral vaccines in the 1950s (particularly for poliomyelitis) that accelerated the design of large-scale bioprocesses for mammalian cells (Kretzmer 2002). These processes required the use of anchorage-dependent cells and the modern version of this viral vaccine technology currently employs micro-carrier support systems that can be used in pseudo-suspension cultures designed in stirred tank bioreactors. Tumor cells can grow anchorage-dependent and anchorage-independent (suspended). Most cells, other than lymphoid cells and some malignant cells, must be attached to a substrate in order to replicate. It was discovered that many cell lines and tissues can grow more rapidly on a collagen matrix than on a glass plate. Subsequently, a great number of cell lines have been cultivated on a collagen matrix (Kleinman et al. 1981). Collagen is also coated on micro-carriers (Gebb et al. 1982) or used as a material for porous micro-carriers (Van Brunt 1986). Advantages of suspension cells are an easier scale-up and the reduction of costs due to micro-carriers. Possible disadvantages include difficulties of cell retention.

A number of immortalized rodent cell lines have been isolated from cultures of normal fibroblasts. Instead of dying as most of their counterparts do, a few cells in these cultures continue proliferating indefinitely. Such permanent cell lines have been particularly useful for many types of experiments because they provide a continuous and uniform source of cells that can be manipulated, cloned, and indefinitely propagated in the laboratory. They are referred to as immortal cell lines. That human tumors could also give rise to a continuous cell line, e.g., HeLa (cervical cancer cell line from patient Henrietta Lacks), encouraged the interest in human cells. The technical potential of cell culture began to reshape genetics in the 1950s and 1960s, when techniques for hybridizing cells seemed to provide a way around the inconvenience of sexual breeding. In 1975, hybrid cells (known as hybridomas) were produced from the fusion of two or more cells capable of continuous production of a single type of antibody and are now produced commercially in kilogram quantities from large-scale cultures of hybridoma cells. Recombinant DNA technology, also known as genetic engineering, was developed in the 1970s to express mammalian genes in bacteria and to produce recombinant proteins. This technology stimulated the enhanced interest in mammalian cell culture bioprocesses (Butler 2005). It soon became apparent that large complex proteins (and especially those having therapeutic value) couldn't be produced in bacteria, as they do not have the appropriate metabolism to add sugar chains (glycosylation) to these proteins to give them their natural activity. Therefore, genetically-engineered animal cells were developed for large-scale commercial production of such important proteins. Animal cells are now becoming the prevailing expression system for the production of recombinant proteins because of their capacity for proper protein folding, assembly, and post-translational modifications.

Hybridoma cells secrete monoclonal antibodies whereas Vero (kidney epithelial cells from African green monkey) are mainly used to produce viruses and Chinese hamster ovary (CHO) cells are usually used in production recombinant proteins including monoclonal antibodies. CHO cell lines are widely used for being highly stable expression systems for heterogonous genes (Wurm 1997, 2004). CHO cells have become the standard mammalian host cells employed in the production of recombinant proteins used in human therapy, such as growth factors, anti-thrombolytic and monoclonal antibodies (Chu and Robinson 2001), although the mouse myeloma (NSO), baby hamster kidney (BHK), human embryonic kidney (HEK-293) or human-retina-derived (PER.C6) cells are alternatives. Also, interest in insect cell culture technology has increased during the last 2 decades. Promising areas of application are the production of insect pathogenic viruses as insecticides for agricultural pest control and the production of recombinant proteins using the baculovirus expression system for biological research and human health care. The use of stem cells and other nontraditional cells for human therapy is becoming an area of increasing importance for media formulation.

The advantage of CHO and NSO cells is that there are well-characterized platform technologies that allow for transfection, amplification and selection of high-producer clones. Transfection of cells – the process of introducing nucleic acids into cells by non-viral methods – with the target gene along with an amplifiable gene, such as dihydrofolate reductase (DHFR) or glutamine synthetase (GS), has offered effective platforms for expression of the required proteins (Butler 2005). High yields of recombinant proteins can also be produced from a human cell line, notably PER.C6. This cell line has been well characterized and has been shown to be able to produce high levels of recombinant protein with relatively low gene copy numbers and without the need for amplification protocols. The added value of these cells is that they ensure the recombinant proteins produced receive a human profile of glycosylation (Butler 2005). Further commercially available adherent cell lines are the epithelium cell line Madin-Darby canine kidney cells (MDCK) as well as designer cell lines like PER.C6, human embryonic kidney (HEK-293), or the duck embryonic stem cells derived AGE1.CR and EB66.

The key biopharmaceutical product groups in biotechnology are: anti-coagulants, blood factors, cytokines, fusion proteins, growth factors, hormones, monoclonal antibodies, nucleic acids, polysaccharide vaccines, recombinant vaccines, and therapeutic enzymes. At the present time, there are up to 30 licensed biopharmaceuticals produced from mammalian cell bioprocesses and over 75 drugs are currently in different phases of their development. These are defined as recombinant proteins, monoclonal antibodies and nucleic acid-based products. Since 1996, the chimeric and humanized monoclonal antibodies have dominated this group with such blockbuster products as Rituxan, Remicade, Synagis and Herceptin (Brekke and Sandie 2003; Pavlou 2004). A chimeric antibody (e.g., Rituxan) consists of a molecular construct in which the mouse variable region is linked to the human constant region. A further step to humanizing an antibody can be made by replacement of the murine framework region, leaving only the complementarity determining regions (CDRs) that are of murine origin. These hybrid construct molecules are far less immunogenic than their murine counterparts and have serum half-lives of up to 20 days (Butler 2005).

Serum- and Animal-Free Medium Adaptations

Traditionally, mammalian cells have generally have been cultivated in a nutrient mixture (basal medium) supplemented with 5–20% of serum or serum-derived components. Fetal calf serum (FCS) also called fetal bovine serum (FBS), is a rich source of a multitude of component substances such as hormones (like insulin), growth factors, transport proteins (such as transferrin and albumin), nutrients, attachment factors, and others. However, serum is also undefined in terms of absolute composition. This may lead to inconsistent growth and productivity. Its high albumin content ensures that the cells are well-protected from potentially adverse conditions such as pH fluctuations or shear forces. Furthermore, the introduction of animal-derived proteins leads to challenges due the need to eliminate these proteins during downstream processing. Concerns in the biopharmaceutical industry over transmissible spongiform encephalopathies (TSEs) and other adventitious agents from using agents arising from the use of animal-derived components have prompted the elimination of animal-derived components and implementation of new supplements. The potential introduction of prions responsible for transmissible spongiform encephalopathy (TSE or BSE in cows) has truly spurred significant movement away from the use of sera. This has led regulators worldwide to ask manufacturers to minimize bovine-components in substances and media used in drug production. Consequently, national and international legislatures are increasingly opposed to the use of animal-derived products such as FBS in biopharma production. For years, both the US Food and Drug Administration (FDA) and the European Medicines Agency (EMA) have tightened up the rules on all animal-derived components used in commercial-scale production media in a bid to further enhance patient safety. These rule changes have had the greatest impact on the biopharmaceutical manufacturing sector, which has traditionally been reliant on growth media containing animal proteins. To avoid ingredients of animal origin, suppliers of media, feed, and ingredients have responded with media using ingredients of plant origin or manufactured through recombinant expression.

The use of serum-free, protein-free and chemical-defined media are gradually becoming one of the goals of cell culture especially for standardizing culture conditions or for simple purification of cell products like monoclonal antibodies. Due to the safety aspects with respect to contamination risk with viruses, prions, etc., there is a strong development to replace animal-derived components by plant-derived components. Furthermore, national as well as international legislation is developing, e.g., by EU and FDA, to ban the use of animal-derived product as much as possible. Consequently, serum and animal components are progressively being from new culture medium formulations. Protein-free media have many advantages, i.e. (i) chemical definition, (ii) reduced cost as compared to media supplemented with recombinant proteins, (iii) facility of purification of recombinant secreted proteins, etc. Unfortunately, their utilization can be associated with some major drawbacks, i.e. decrease in cell growth, decrease in protein secretion, increase in proteolytic risk by the lack of serum protease inhibitors, etc. Protein hydrolysates and particularly plant-derived protein hydrolysates provide some of the solutions to these limitations. Protein hydrolysates can increase both cell growth rate and recombinant protein secretion and some proteases could act as a source of inhibitors by providing competitive substrates, but they decrease the chemical definition of the media. Protein hydrolysates (peptones) have been used to replace serum and serum components in a variety of mammalian cell culture processes since the 50ts. Depending on their origin and the degree of hydrolysis, these protein hydrolysates may supply nutrients, adhesion components or growth factor analogues (Nyberg et al. 1999; Heidemann et al. 2000). Lactoalbumin hydrolysate was a good amino acid and nutrient source of serum-free media at an early stage of the research. Primatone™ RL (Sheffield Bio-Science™) has successfully been used to reduce or replace fetal calf/bovine serum in human, insect and rodent large-scale cell culture productions. Primatone™ RL, a water-soluble enzymatic digest of animal tissue, is a low-cost medium supplement of a complex nature which serves as a source of amino acids, oligopeptides, iron, some lipids and other races of low molecular weights. In some cases, the combination of two protein hydrolysates from different protein sources produced significantly greater recombinant protein yields than a single peptone by increasing growth and/or specific productivity. Plant protein hydrolysates, particularly cottonseed hydrolysates such as HyPep™ 7504 or UltraPep™ Cottonseed, have shown to be good alternatives to the use of animal products. They are obtained after enzymatic hydrolysis of plant proteins and consist of a mixture of oligopeptides, free amino acids, and carbohydrates. Plant protein hydrolysates are now widely used as key medium additives in serum-free cell culture processes for industrial production of therapeutic recombinant proteins.

Cellular Metabolism, Tumor Cells in Culture, and the Influence of Protein Hydrolysates

Protein hydrolysates are commonly utilized in cell culture process as a component of a complete medium formulation or as part of a fed-batch bioreactor process. Nyberg et al. (1999) demonstrated that significant amounts of amino acids are liberated from peptides during cultivation of CHO cells in a serum-free medium containing the animal tissue hydrolysate Primatone™ RL (Sheffield Bio-Science™ Pharma Ingredients). It is well documented that protein hydrolysates can have a substantial positive impact on cell growth and/or protein production by the cell. If cells are grown on free amino acids, a high intracellular pool of free amino acids is found as the accumulation of free amino acids is determined by the free amino transport system. However, if the cell growth in culture is based upon oligopeptides, then the intracellular pool of free amino acids is determined by the peptidase activity of the cell itself. This likely affects overall cell metabolism. Another explanation is that the uptake of intact peptides followed by intracellular hydrolysis represents an energetically more efficient mechanism than the uptake of free amino acids (Daniel et al. 1992). In this way, more amino acids can be used for growth instead being just a source of energy. In theory, the uptake of intact peptides followed by intracellular hydrolysis represents an energetically more efficient mechanism than the uptake of free amino acids. To figure this out, Bonarius et al. (1996) estimated the intracellular metabolic fluxes in hybridoma cells in a continuous culture. The uptake and production rates of amino acids, glucose, lactate, O_2, CO_2, NH_4^+, MAB (monoclonal antibodies), and the intracellular amino acid pools were estimated for a medium with ('optimal medium') and without ('suboptimal medium') Primatone™ RL (Sheffield Bio-Science™), an enzymatic hydrolysate of animal tissue. A more than two fold increase in cell density and some 50% extra MAB production was found. The overall amino acid composition was the same for both media. Bonarius et al. (1996) found that in a cell culture medium with Primatone™ RL, the flux through the pentose phosphate shunt is higher than in a suboptimal medium without Primatone™ RL (Fig. 1). As the pentose phosphate pathway serves to generate NADPH to stimulate anabolism, it means more synthesis of cellular products. Bonarius et al. (1996) also found that the entry of pyruvate in the TCA cycle via pyruvate dehydrogenase (as acetyl CoA) in culture medium with and without Primatone™ RL was respectively 8% and 14% of the glucose uptake rate. Thus, when Primatone™ RL is added to the culture media, more glucose is directed to the TCA cycle than to glycolysis. In other words,

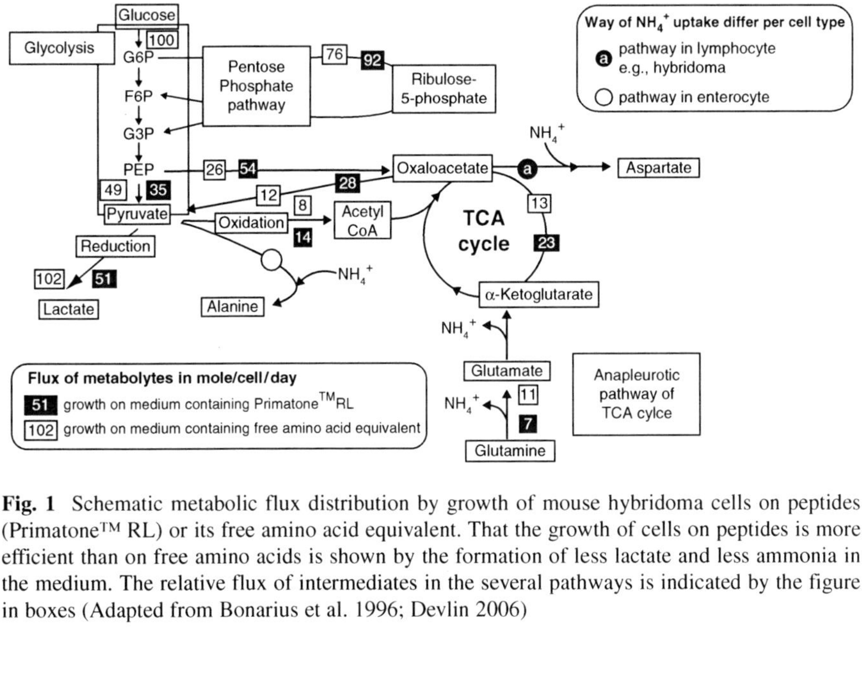

Fig. 1 Schematic metabolic flux distribution by growth of mouse hybridoma cells on peptides (Primatone™ RL) or its free amino acid equivalent. That the growth of cells on peptides is more efficient than on free amino acids is shown by the formation of less lactate and less ammonia in the medium. The relative flux of intermediates in the several pathways is indicated by the figure in boxes (Adapted from Bonarius et al. 1996; Devlin 2006)

Primatone™ RL stimulates a more energy-efficient metabolism compared with the equivalent free amino acid mixture. A higher respiration quotient (RQ) in the medium with Primatone™ RL was also found which was explained by the concomitant increased fatty acid synthesis rates and with high ratios of glucose/ glutamine oxidation (Bonarius et al. 1996). This means more glucose oxidation and less glutamine oxidation which leads to less lactate and less ammonia production respectively. Overall, it means that by the addition of Primatone™ RL leads to less glycolysis and glutaminolysis, and in this way, a more energy efficient cell metabolism.

Enhanced Production by Cultured Cells on Plant-Derived Protein Hydrolysates

Because of safety concerns related to the use of animal-derived products in processes producing biologics for human injection, plant-derived hydrolysates are commonly favored over animal tissue hydrolysates. Regardless of the source of the material, the major nutrients that protein hydrolysates provide to the culture media include free amino acids, peptides, carbohydrates, vitamins, minerals, and undefined components. By far the most important compounds from a mass-balancing viewpoint that hydrolysates supply are amino acids and peptides. Some hypotheses have been formulated about the possible roles of protein hydrolysates. Jan et al. (1994), Schlaeger (1996), and Heidemann et al. (2000) have hypothesized that mixtures of various low-molecular weight (MW) peptides supply the cells with amino acids that may mimic an optimized free amino acid content of the protein-free medium. Nevertheless, Rasmussen et al. (1998) concluded that peptides in the plant hydrolysates could function as growth factors for CHO cells. Heidemann et al. (2000) also suggested a potential effect of higher-MW oligopeptides as growth or survival factors.

Low protein and protein-free media fortified with plant-derived protein hydrolysates can efficiently support the growth and productivity of CHO cell lines in suspension and on micro carriers in the absence of serum (Verhoeye et al. 2001). The growth of CHO 320 cells, a clone of CHO K1 cells genetically modified to secrete human interferon-γ (IFN-γ), adapted to the culture in suspension, and the production of IFN-γ are significantly better in the serum-free medium in presence of a rice protein hydrolysate (Sheffield™ Bio-Science) in its absence. Both cell growth and IFN-γ secretion were found to be equivalent to those reached in serum-containing medium (Bare et al. 2001).

Several plant protein hydrolysates, selected on the basis of their content of free amino acids and oligopeptides with different molecular weights ranging from 1 to 10 kDa and a small proportion of peptides higher than 10 kDa, were tested for their ability to improve culture parameters. Most of the beneficial effects of protein hydrolysates correlated with their content of larger oligopeptides that could selectively be enriched in sequences that might mimic particular animal proteins. However, if the protein hydrolysate supplies peptides with specific sequences acting as growth, survival, or protecting factors, it could improve the performance of the medium in this way. These protein hydrolysates do not seem to add significantly to the nutritive potential of basal protein-free nutritive medium. The results suggest that the use of plant protein hydrolysates with potential growth factor-like or anti-apoptotic bioactivities could improve mammalian cell cultivation in protein-free media while increasing product biosafety (Burteau et al. 2003). This is in agreement with evidence of oligopeptide effects on cultured animal cells (Franek and Katinger 2002; Franek et al. 2003). Burteau et al. (2003) reported that several plant-derived protein hydrolysates could improve cell growth and production. For instance, a wheat gluten hydrolysate (HyPep™ 4605 with about 1% free amino acids, Sheffield™ Bio-Science) improved the growth of CHO-320 cells in a shake-flask experiment up to 30% and recombinant human interferon-γ (IFN-γ) secretion up to 60%, which was even better than that reached in media containing 10% (v/v) serum. The maximal density of viable cells increased as a function of the concentration of the wheat protein hydrolysate. Significant responses were found during both the exponential and the stationary phases. No such positive effect was observed when the wheat protein hydrolysate was replaced by a mixture of various nutrients. This also suggests that the effect of the protein hydrolysate is probably not entirely nutritional (Burteau et al. 2003). The macro heterogeneity, especially important with respect to the glycoforms of the recombinant IFN-γ produced, was unaffected by the presence of the wheat protein hydrolysate. The glucose, glutamine, and alanine profiles were similar for all the conditions, but the final concentration of ammonia reached about 6 mM (0.25 mg/mL) in the absence of protein hydrolysate (or with 0.25 mg protein hydrolysate per mL), but rose to 7 m*M* with either 1 or 2 mg protein hydrolysate per mL. Lactate concentration decreased from 28 m*M* (72 h) to 24 m*M* in PF-BDM (Burteau et al. 2003), confirming its consumption after glucose depletion, but the lactate concentration decreased from 28 to 18 m*M* in PF-BDM supplemented with 2 mg protein hydrolysate per mL. A slightly decreased content of amino acids in the culture medium was shown during the exponential and the stationary phases and an increase at the end of the batch, only when medium was fortified by plant protein hydrolysate. As suggested by Heidemann et al. (2000), peptones are partially taken up and cleaved into free amino acids at the plasma membrane or inside the CHO 320 cells (or both) by various proteases. During the death phase, the produced free amino acids, which are no longer consumed, would accumulate in the culture medium, consequently increasing the final free amino acid content (Burteau et al. 2003). The addition of Peptose (Tryptose, Oxoid) to a hybridoma WuT3 culture stimulated the specific antibody production and the authors suggested that the protein hydrolysate raised the efficiency of glycolysis and promoted the utilization of amino acids up to 5 g protein hydrolysate per liter (Zhang et al. 1994).

Thus, plant-derived protein hydrolysates can successfully replace animal proteins in the formulation of media for animal cell culture (Franěk et al. 2000; Heidemann et al. 2000; Burteau et al. 2003; Sung et al. 2004; Chun et al. 2007). In the presence of various plant-derived protein hydrolysates, wild type CHO-K1 cells and recombinant

CHO-320 cells demonstrated improved cell performances in terms of cell growth and productivity whereas the glycosylation pattern of the secreted glycoproteins was not altered. Furthermore, the presence of protein hydrolysates allowed a more rapid adaptation of CHO cells to serum-free and protein-free media and reduced cell death occurring during the transition phase. Finally, it was demonstrated that the addition of plant-derived protein hydrolysates to cryopreservation solutions for animal cells improves cell survival during the freezing and thawing steps (Verhoeye et al. 2001). The successful use of plant protein hydrolysates in embryo freezing medium also has been demonstrated (George et al. 2006).

Development of Plant-Derived Protein Hydrolysates for Cell Culture Media

The following are important and should be seriously considered when developing and using protein hydrolysates. Are peptides used:

(i) Directly as amino acid source (~ replacement of free amino acids)
(ii) Indirectly as a stimulator (~ serum replacement) of the growth (~biomass) of mammalian cells or
(iii) Production (~biopharmaceutical) by those cells, and
(iv) Protection of cultured cells against shear stress (~apoptosis)

When peptides are used as amino acid source, then the amount of free amino acids and short chain peptides might be important as di- and tri-peptides can be directly transported through the cell membrane where the peptides in the cytosol or lysosomes will be further digested into free amino acids. In that case, the amount of free amino acids and the small peptides is very important. If, however, peptides stimulate the growth in a more indirect way, then the conformation of the peptides and thus their amino acid sequences (~3D configuration) are important.

Kerry Group with its Sheffield™ Bio-Science products was among the first to market non-animal-derived protein hydrolysates for both serum replacement and general media/system optimization efforts (Blom et al. 1998). In the mid-1990s, experiments were conducted by us to understand if plant protein hydrolysates could replace the amino acid mixtures in standard cell culture media or whether protein hydrolysates had other functionalities. To find out if free amino acids and di- and tri-peptides were important, hydrolysates were developed with different amounts of free amino acids and different peptide distributions. Wheat gluten was selected as a protein source to start with as the amino acid composition resembles closely the free amino acid mixture of standard cell culture media. With hybridoma cells (anti-CD20 mouse lymphoma) it was found that a supplement of a wheat gluten HyPep™ hydrolysate, containing about 3.5% free amino acids, and 6% di- and tripeptides, specifically stimulated the production of monoclonal antibodies (IgG) (Fig. 2, internal communication). The control media was standard RPMI 1640 with the free amino acid mixture. The experimental medium with the HyPep™ was supplemented

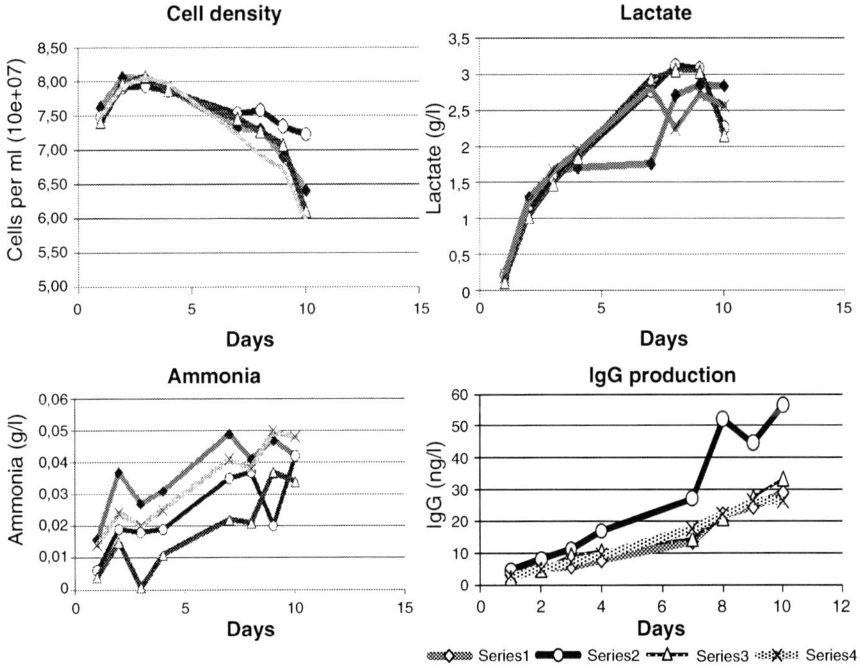

Fig. 2 T-flask experiments. Series 1: RPMI 1640 with free amino acid mixture (control); Series 2: RPMI 1640 plus 1.07 g/L Wheat HyPep™; Series 3: RPMI 1640 plus 0.535 g/L Wheat HyPep™; Series 4: RPMI 1640 plus 0.535 g/L Wheat HyPep™ plus 2 mM glutamine. Media contained 8% FBS. The wheat HyPep™ highly stimulated IgG production with no influence on cell growth or lactate compared with the control; ammonia production was slightly less

with a mix of free amino acids to equalize the amino acid composition of both control and experimental media with the HyPep™. Both media were supplemented with 8% bovine serum. It was found that although the hybridoma cells grew almost at the same rate as on the control media, the cells stayed vital for a longer period of time when the media contained about 1 g HyPep™ per liter. Also the hybridoma cells on the HyPep™ media significantly produced more monoclonal antibodies (IgG) and excreted less ammonia whereas the lactate production was comparable. Reducing the amount of HyPep™ resulted in less IgG production in proportion to the lower amount of the HyPep™ (more details in Fig. 2). Under these conditions, even in the presence of serum, wheat gluten hydrolysate stimulated monoclonal antibody production without stimulating cell density (~biomass). The hybridoma cells seem capable of producing more products without increasing glutaminolysis as no extra (and perhaps even less) ammonia was excreted in the culture media. This suggests that the TCA cycle worked more efficiently when the HyPep™ was added. It appears like a less truncated TCA cycle is present in these hybridoma cells. Lactate production was more or less the same in all cell culture media and seems to result more from the glucose concentration in the media. This was also observed by Bonarius et al. (1996) who mentioned that tumor cells have lost control of glucose uptake and therefore the extracellular concentration of glucose mainly determines the glycolytic flux in the tumor cell.

The addition of at least 1 g/L HyPep™ stimulated the IgG production without producing more ammonia. It therefore appears more likely that the protein hydrolysates stimulate a more efficient use of amino acids by hybridomas, as has been suggested before (Zhang et al. 1994). Noteworthy is the observation that even in the presence of serum, the wheat protein hydrolysate positively influenced antibody (IgG) production in these hybridoma cells. Thus the protein hydrolysate offers functionality above that offered already by serum. That the oligopeptide fraction remains almost unaltered during culturing, as we found, suggests a role for protein hydrolysates other than a nutritional one. Because larger peptides mainly seem to stimulate the growth and/or production in a more indirect way, the conformation of these peptides and thus their amino acid sequence (~3D configuration) might be very important. To test this further, different types of peptides were developed by combinations of plant proteins from different sources including wheat, soy, pea and cotton, and different non-animal derived enzymes. By using combinations of different enzyme(s) and different downstream processing, the wheat gluten hydrolysates HyPep™ 4601 (~5% free amino acids), HyPep™ 4602 (~30% free amino acids), HyPep™ 4603 (~40% free amino acids) and HyPep™ 4605 (~1% free amino acids) were developed. These hydrolysates were intensively tested in cell culture media under serum-free conditions in several types of mammalian cell lines including CHO, BHK, VERO cells and several Hybridoma cell lines. It was clearly shown that these wheat gluten hydrolysates indeed can replace serum very well when added in combination with 1% ITS (Insulin-Transferrin-Selenium). Especially in spinner tests, HyPep™ 4605 supplemented to standard DMEMF/F12 medium including the standard amino acid mixture and added ITS, highly stimulated the growth of hybridoma cells and production of IgG under serum-free conditions. In T-flasks it was shown that the peptides in HyPep™ 4602 better promote growth and production (IgG) than the peptides of HyPep™ 4601. The experiments not only show that plant protein hydrolysates in serum-free media stimulate the growth and performance of mammalian cells, but also offers the cells protection under stress conditions in spinner tests. Especially HyPep™ with larger peptides (wheat gluten-based HyPep™ 4605, rice based HyPep™ 5603, the pea-based HyPep™ 7401, the soy-based HyPep™ 1510 and the cotton-based HyPep™ 7504) offer shear stress protection to the cells in a way comparable with albumin. It makes the cells more robust and more resistant to shear stress. These protein hydrolysates furthermore have a function in attachment and spreading of attached cell lines like Vero. Also, the cotton-derived HyPep™ 7504 (~13% free amino acids) and the soy-derived HyPep™ 1510 (~8% free amino acids) highly stimulate growth and production in spinner tests with hybridoma cells, while HyPep™ 4605 produced the best growth and IgG production in this particular test (Fig. 3).

The optimal concentration of HyPep™ products is dependent upon the cell line in culture and the experimental conditions. This was true for the CHO cell line when experiments were performed using T-25 flasks or spinner flasks. For the Vero

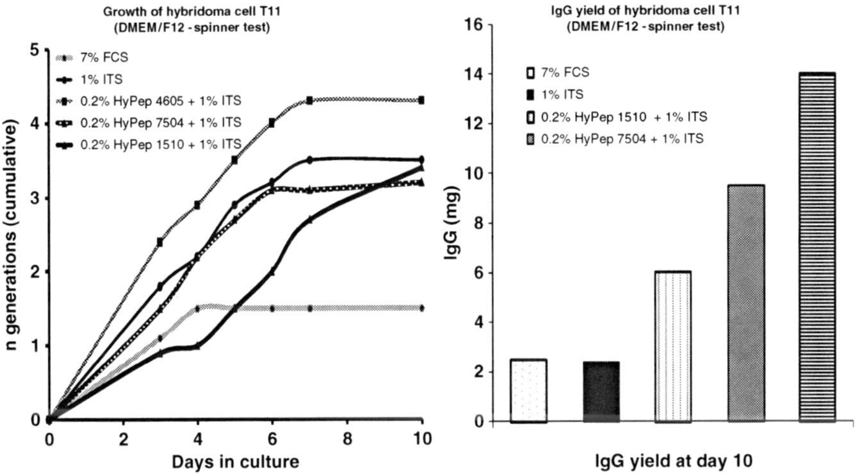

Fig. 3 Growth and IgG production of cultured hybridoma cell T11 in a spinner test in standard DMEM/F12 media (Hyclone; UT). HyPep™ 1510 is soy-based, HyPep™ 4605 is wheat gluten-based and HyPep™ 7504 is cotton based. The growth and production in this test is specifically stimulated by the addition of HyPep™ 4605

cell line, the HyPep™ 4602 was found to be beneficial for the growth and for the attachment of the cells to the growth surface (the effect was also seen for CHO cells growing as attached cells, using micro carriers). It seems that the HyPep™ products containing the larger peptides are beneficial for attached cells whereas no clear effect is seen when used for suspension cells. Vero cell lines seem to benefit from the addition of the HyPep™ products containing only a few percent of free amino acids. The cells did attach better and the growth performance of the Vero cells after adaptation also went up compared to the control (cells growing in medium without HyPep™ products). The morphology of the cells is better; the growth rate is faster, etc.

For a better understanding of the importance of the different fractions with respect to peptide distribution, fractions were made from HyPep™ 4601 and HyPep™ 4605, by using dialysis tubes with different molecular weight cut-offs of 3.5 and 1.0 kDa. Both dialysates (the permeates and retentates) were evaporated and freeze-dried. The powders were tested on Hybridoma cells, Vero and on CHO cells. The tests clearly showed growth stimulation by the larger molecular weight fractions whereas the smaller molecular weight fractions had no growth promoting effect. Furthermore, it became clear that hydrolysates with a large amount of free amino acids may interfere with the amino acid composition of the standard basic medium. This should be taken into account when developing media for a specific cell line.

Although the process of manufacturing is well controlled, the nature of protein hydrolysates are not fully defined and batch-to-batch variations are known to exist. Therefore, Sheffield™ Bio-Science developed a new hydrolysate platform that addresses the need for animal-free cell culture medium supplements while also minimizing variability concerns. The platform is based upon a novel approach to enzymatic digestion and more refined processing. At the heart of the product is a rationally designed non-animal enzyme cocktail, including both proteases and non-proteolytic hydrolases, whose activities can also liberate primary components of the polymerized non-protein portion of the raw material. This enzyme blend is added during a highly optimized process step that targets specific enzyme-substrate reactions to expand the range of beneficial nutritional factors made available to cells in culture. Such factors are fundamental to improving the bio performance of the culture system, as they provide not merely growth-promoting peptides and amino acids, but also key carbohydrates, lipids, minerals, and vitamins. In this way, both rate and quality of protein expression are improved as well as culture life due to the enhanced osmo-protectant and anti-apoptotic properties. Compared to a typical enzymatic digestion, this highly optimized digestion process requires fewer steps and intrinsically yields a better-controlled and therefore a more reproducible product and reduces or eliminates many of the root causes that may lead to inter-lot variability of the finished product and subsequent inconsistency in end-use performance. This more sophisticated production process yields hydrolysates that are more amenable to sterile filtration, precluding the need for processing adjuncts or additives. This allows hydrolysate end users to experience streamlined media

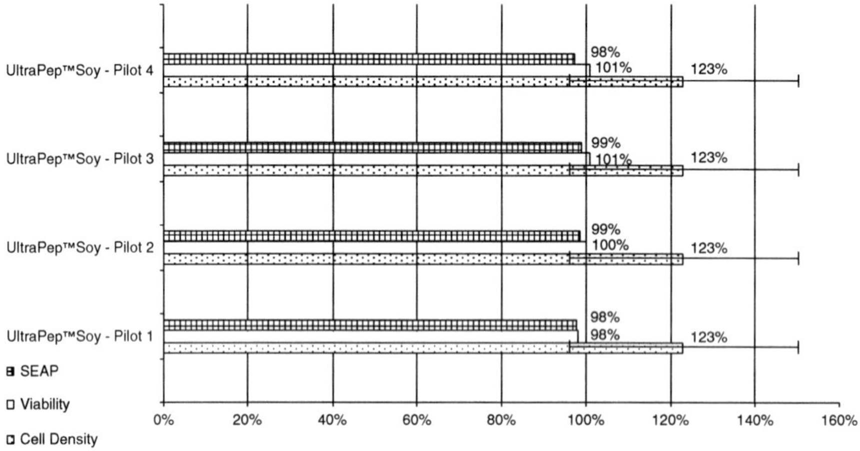

Fig. 4 UltraPep Soy performance in Sheffield™ Bio-Science Clone B.1 cells, a transfected CHO-K1. Pilot lot performance of Sheffield™ Bio-Science Clone B.1 in culture as a percentage of Hy-Soy™. Sheffield™ Bio-Science Clone B.1 is a transfected CHO-K1 line engineered to constitutively express secreted embryonic alkaline phosphatase (SEAP) by means of a modified human cytomegalovirus (HCMV) promoter. Monolayer cultures were grown in six-well microplates containing a final medium volume of 3 mL/well. The basal medium consisted of 50% chemically defined medium (CDM4CHO™, HyClone®, Logan, UT) and 50% Ham's F12-K (HyClone®, Logan, UT), 1 mg/mL G-418 (Mediatech, Inc., Manassas, VA), supplemented with 5% fetal bovine serum (FBS)(HyClone®, Logan, UT). Cultures were seeded at 2 × 10e5 cells per well, and incubated at 37°C in 5% CO2. Data were collected after 5 days

preparation and bioreactor supplementation activities. The first of these new types of hydrolysates, UltraPep™ Soy, was introduced to the market in 2006, while and UltraPep™ Cottonseed was introduced in 2009. The data obtained with UltraPep™ platform is consistent and has been presented at multiple professional conferences, including IBC (12–16 Oct 2009, Raleigh, USA) and ESACT (16–20 June 2007, Dresden, Germany and 07–10 June 2009, Dublin, Ireland). A sampling of these data is shown in Fig. 4.

Discussion

Protein hydrolysates for pharmaceutical applications have at least three functions. Peptides in the hydrolysate are used (i) directly as an amino acid source (replacement of free amino acids), (ii) indirectly as a stimulator of growth and/or production, and (iii) protection of cells against shear stress. For the replacement of free amino acids in the culture medium, this implies that the amount of free amino acids and di- and tripeptides are important in the protein hydrolysate. For the stimulation of growth and/or production as well as protection against shear stress, however, the conformation of the peptides, and thus its amino acid sequence (~3D configuration), is more important. Several lines of evidence suggest that peptides in plant hydrolysates can act as a kind of growth factor and/or production stimulator for CHO cells and hybridoma cells rather than as nutrients (peptides used as an amino acid source). The addition of longer peptides in HyPep™ to serum-free culture media makes CHO and hybridoma cells more robust in spinner flasks and at least the hybridoma cells produce more antibody in comparison with the control at comparable growth rate. The positive effect of the larger peptides is due to stabilizing cell membranes and/or providing exogenous growth factors (mitogen and/or viability factor). The tests also suggest that peptides can stimulate production by the cells, which is not necessarily parallel to the growth rate. Additionally, the non-protein components within the hydrolysates seem to play positive roles in both growth and product quality. The beneficial effects of the plant protein hydrolysates of the HyPep™ series on the industrially most important cell types are currently well recognized (Table 1). These include various hybridoma lines, African green monkey kidney cells (VERO), Chinese hamster ovary cells (CHO), baby hamster kidney cells (BHK), *Spodoptera frugiperda* insect cells (*Sf9*) infected by the baculovirus, and HEK (human embryonic kidney).

Protein hydrolysates with larger peptides in the cell culture media seem to bring about more efficient cell metabolism with a more efficient use of amino acids, at least in hybridoma and CHO cells. Noteworthy is that even in the presence of serum, wheat protein hydrolysate can positively influence the antibody (IgG) production by hybridoma cells. Protein hydrolysates therefore offer a functionality that is above what is offered already by serum. Furthermore, the oligopeptide faction remains almost unaltered during culturing, which suggests another role for protein hydrolysates than a nutritional one. The effects might suggest

Table 1 A sampling of Published US Patents/Applications by Major BioPharmaceutical companies (As of June 18, 2009)

Assignee	Hydrolysate	Drug entity/Description	Host cell	Patent	Year
Genentech	Primatone	IGF-1R	Insect cells	20090068110	2009
Astra Zeneca	HyPep 1510	IgE	HEK-293	20090060919	2009
Polymun Scientific	HyPep 1510	Serum free media/Influenza vaccine	VERO	7494659	2009
Invitrogen	HyPeps, Hy-Yest 444	Serum Free Media	VERO	20090061516	2009
Baxter	HyPep 1510	AOF media	Multiple	20080064105	2008
Merck	HyPep 1510, HyPep 4601	Fc-EPO fusion protein	BHK	7465447	2008
Amgen	Soy peptone	Fab antibody (OPGL)	CHO	7364736	2008
Immunex/Amgen	N-Z-Soy, Hy-Soy	Fusion Proteins (rTNF)	CHO	7294481	2007
Genentech	Primatone	P37 rProtein (ADAM8)	CHO	7226596	2007
Novo Nordisk	HyPep 4601, HyPep 4605	Factor VII	BHK	20060216790	2006
BASF	HyPep, Edamin S	Serum Free Media	Insect Cells	6670184	2003
Genentech	Primatone	Glycoprotein	CHO	6610516	2005
Chugai	HyPep 4402, HyPep 4601	Serum Free Media	CHO	6406909	2002
Wyeth/AHP	N-Z-Amine A	Vaccine stabilization	VERO	5958423/09	1999
Roche	Primatone	anti-PGDF	CHO	5882644/03	1999
Genzyme	Primatone	Serum free media	Hybridoma	56 91202/11	1997

involvement of transduction proteins in the membrane, such as G protein-coupled receptor, or other transducers and causes an activation of some enzymatic function on the cytoplasmic side of the membrane. In this view, peptides can hormonally affect (stimulate or inhibit) cell metabolism. Peptides can stimulate the growth of the cell by interaction with a kind of receptor on the cell surface, by specific peptides functioning as a carrier (transporter/detoxifier) for instance of lipids, metal ions or other toxic substances, or by stabilizing the fragile cell membrane by interaction with larger peptides. In this view, the hydrophobic and/or the hydrophilic properties of peptides (surface) become very important. Therefore, the physical-chemical character related to the sequence of amino acids in the peptides is very important.

Comparable information was suggested by the work of Franěk et al. (2000) who showed that the addition of synthetic peptides to protein-free medium generally resulted in an improvement of at least one of the hybridoma cell culture parameters: long-term viability, growth rate, or productivity rate. Recognizing the beneficial effects of the plant protein hydrolysates on animal cell cultures, this group tried to identify the type of peptides responsible for growth stimulation of hybridoma and CHO cells. They found that the enhancement of monoclonal antibody production not was correlated with the promotion of cell growth. Growth- and production-enhancing effects of peptides increase with peptide-chain length and at a concentration of >1 mM. It was further found that peptides have anti-apoptotic activity and that some peptides promote growth while other peptides suppress growth but enhance product yield. The nutritional use of amino acids liberated from the peptides was, at most, marginal. The addition of longer peptides containing HyPep™ to serum-free culture media can make CHO and hybridoma cells more robust in spinner flask experiments. This was true for gluten-based HyPep™, soy-based HyPep™ and the cottonseed-based HyPep™. This suggests that specific plant hydrolysates from the HyPep™ series can fulfill "albumin-type" functions in lowering the shear stress and making the cells more robust. Rice-based HyPep™ has interesting properties for the adaptation of CHO cells to serum-free medium and reduces cell death occurring during the transition phase. Soy based HyPeps™ further highly stimulate production by CHO cells.

In conclusion, experiments with several mammalian cells demonstrated an improvement of cell performance in terms of cell growth and productivity in the presence of various plant-derived protein hydrolysates from the HyPep™ series. Also glycosylation pattern of the secreted glycoproteins, i.e., IFN-γ by CHO cells, were not altered in HyPep™-fortified culture media in comparison with serum-supplemented media. It was further found that peptides have anti-apoptotic activity and that some peptides promote growth while other peptides might suppress growth but specifically enhance product yield. It was also demonstrated that the addition of HyPep™ to cryopreservation solutions for animal cells improves cell survival during the freezing and thawing steps. Several HyPep™ products demonstrate adhesive, fibronectin-like, activity as they promote the attachment and subsequent spreading of many cells. It must be stressed that every cell line might have a preference for a specific HyPep™.

Summary and Outlook

A strong trend has emerged to remove not only serum, but also any product of animal origin, from animal cell culture media during the production of recombinant proteins. This shift will improve biosafety and drive next generation of downstream processing technologies. In the absence of serum, the biopharmaceutical industry has embraced the supplementation of culture media with plant-derived protein hydrolysates and yeast extracts. The beneficial effects of plant-derived hydrolysates on the most industrially-important cell types are currently well recognized. These include hybridoma, African green monkey kidney(VERO), Chinese Hamster Ovary (CHO), Human Embryonic Kidney (HEK), Baby Hamster Kidney (BHK) and baculovirus-infected *Spodoptera frugiperda* insect cells (*Sf9*). With the commercial availability of plant-derived protein hydrolysates (peptones of soy, rice, wheat gluten, pea, and cottonseed), they have become the unanimous choice to supplement basal medium. Protein hydrolysates seem to stimulate a more efficient cell metabolism with a more efficient use of amino acids, often improving cellular metabolism above that produced with serum. Furthermore, the oligopeptide fraction remains largely intact during culturing, which also suggests a role for protein hydrolysates beyond that of a nutrient source.

Since the start of the production of protein hydrolysates more than 60 years ago, almost 20,000 metric tons have found the way through many applications in nutrition, cosmetic, feed, agriculture, diagnostics, fermentation, infant formula and bio pharmaceutical industries. With more than 80 producers in the world, protein hydrolysates are known for their nutritional and functional value and are very well accepted ingredients in different industries. The cosmetic, feed, agriculture and bulk fermentation industries are typically interested in the cost effectiveness whereas the bio pharmaceutical industry on the other hand values partnership and innovation most. The discovery of BSE and other contagious risks from animal origin protein hydrolysates as well as increasing regulatory requirements from the FDA drove the trend toward simple and defined ingredients. For many pharmaceutical and fermentation companies, one of the most important cost factors is improving the yield of their production systems. Some protein hydrolysate producers, like Sheffield™ Bio-Science, producers are actively improving their product portfolios to provide these benefits. New protein hydrolysates will definitively be of animal origin-free, better characterized and better controlled during and after production.

More research needs to be carried out to better understand the functionality of protein hydrolysates. For some applications, peptide fragments have been identified that contain a part of the functionality. It will be an enormous task to conduct research to determine all functionality aspects and the synergistic effects before the effects of protein hydrolysates in all applications are fully understood. This is where the collaboration between protein hydrolysate manufacturers and end users is useful and helps both the parties. The value addition will be much higher when all of the functional components are identified and can be produced consistently batch after batch in an economically viable way. Even without complete understanding of the mechanism still protein hydrolysates are playing a key role in cell culture media today and will only get better as the mechanism of action unfolds.

References

Bare G, Charlier H, De Nijs L, Verhoeye F, Schneider Y-J, Agathos S, Thonart P (2001) Effects of a rice protein hydrolysate on growth of CHO cells and production of human interferon-γ in a serum-free medium. In: Linder-Olsson E et al (eds) Animal cell technology: from target to market. Kluwer, Dordrecht, pp 217–219

Blom WR, Kunst,A, Van Schie BJ, Luli GW (1998) Method for *in vitro* cell growth of eucaryotic cells using low molecular weight peptides. U.S. Patent No. 5,741,705. U.S. Patent and Trademark Office, Washington, DC

Bonarius HPJ, Hatzimanikatis V, Meesters KPH, de Gooijer CD, Schmid G, Tramper J (1996) Metabolic flux analysis of hybridoma cells in different culture media using mass balances. Biotechnol Bioeng 50:299–318

Brekke OH, Sandlie I (2003) Therapeutic antibodies for human diseases at the dawn of the twenty-first century. Nat Rev Drug Discov 2(1):52

Burteau CC, Verhoeye F, Mols JF, Ballez J-S, Agathos SN, Schneider Y-J (2003) Fortification of a protein-free cell culture medium with plant peptones improves cultivation and productivity of an interferon-gamma-producing CHO cell line. In Vitro Cell Dev Biol Anim 39(7):291–296

Butler M (2005) Animal cell cultures: recent achievements and perspectives in the production of biopharmaceuticals. Appl Microbiol Biotechnol 68(3):283–291

Carney DN, Bunn PA, Gazdar AF, Pagan JA, Minna JD (1981) Selective growth in serum-free hormone-supplemented medium of tumor cells obtained by biopsy from patients with small cell carcinoma of the lung. Proc Natl Acad Sci USA 78(5):3185–3189

Chu L, Robinson DK (2001) Industrial choices for protein production by large-scale cell culture. Curr Opin Biotechnol 12(2):180–187

Chun BH, Kim JH, Lee HJ, Chung N (2007) Usability of size-excluded fractions of soy protein hydrolysates for growth and viability of chinese hamster ovary cells in protein-free suspension culture. Bioresource Technology 98(5):1000

Daniel H, Morse EL, Adibi SA (1992) Determinants of substrate affinity for the oligopeptide/H+ symporter in the renal brush border membrane. J Biol Chem 267(14): 9565–9573

Derouazi M, Girard P, Van Tilborgh F, Iglesias K, Muller N, Bertschinger M, Wurm F (2004) Serum-free large-scale transient transfection of CHO cells. Biotechnol Bioeng 87(4):537–545

Devlin TM (2006) Textbook of biochemistry with clinical correlations, 6th edn. Wiley-Liss, New York

Dulbecco R, Freeman G (1959) Plaque production by the polyoma virus. Virology 8(3):396–397

Food and Drug Administration (1997) Points to consider in the manufacture and testing of monoclonal antibody products for human use. (Docket No. 94D-0259). Center for Biologics Evaluation and Research, Rockville, MD

Franěk F, Eckschlager T, Katinger H (2003) Enhancement of monoclonal antibody production by lysine-containing peptides. Biotechnology Progress 19(1):169–174

Franěk F, Hohenwarter O, Katinger H (2000) Plant protein hydrolysates: preparation of defined peptide fractions promoting growth and production in animal cells cultures. Biotechnol Prog 16(5):688–692

Franěk F, Katinger H (2002) Specific effects of synthetic oligopeptides on cultured animal cells. Biotechnology Progress 18(1):155–158

Gebb C, Clark JM, Hirtenstein MD, Lindgren G, Lindskog U, Lundgren B, Vretblad, P (1981) Alternative surfaces for microcarrier culture of animal cells. Developments in Biological Standardization 50:93–102

George F, Vrancken M, Verhaeghe B, Verhoeye F, Schneider YJ, Massip A, et al. (2006) Freezing of in vitro produced bovine embryos in animal protein-free medium containing vegetal peptones. Theriogenology 66(5):1381

Ham RG (1962) Clonal growth of diploid Chinese hamster cells in a synthetic medium supplemented with purified protein fractions. Exp Cell Res 28(3):489–500

Hayashi I, Sato GH (1976) Replacement of serum by hormones permits growth of cells in a defined medium. Nature 259(5539):132–134

Heidemann R, Zhang C, Qi H, Rule JL, Rozales C, Park S, et al. (2000) The use of peptones as medium additives for the production of a recombinant therapeutic protein in high density perfusion cultures of mammalian cells. Cytotechnology 32(2):157

Iscove NN, Melchers F (1978) Complete replacement of serum by albumin, transferrin, and soybean lipid in cultures of lipopolysaccharide-reactive b-lymphocytes. J Exp Med 147(3):923–933

Jan DC, Jones SJ, Emery AN, Al-Rubeai M (1994) Peptone, a low-cost growth-promoting nutrient for intensive animal cell culture. Cytotechnology 16(1):17–26

Kleinman HK, Klebe RJ, Martin GR (1981) Role of collagenous matrices in the adhesion and growth of cells. J Cell Biol 88(3):473–485

Koch R (1876) Untersuchungen ueber bakterien V. Die aetiologie der milzbrand-krankheit, begruendent auf die entwicklungsgeschichte des Bacillus anthracis. Beitr z Biol D Pflanzen 2:277–310

Kretzmer G (2002) Industrial processes with animal cells. Appl Microbiol Biotechnol 59(2–3): 135–142

Loeffler F (1884) Utersuchung uber die bedeutung der mikroorganismen fir die entstehung der diptherie beim menschen, bei der taube und beim kalbe. Mitth a d kaiserl Gesundheitsampte 2:421–499

Mizrahi A (1977) Primatone® in mammalian cell culture media. Biotechnol Bioeng 19:1557–1561

Murakami H, Musui H, Sato GH (1982) Growth of hybridoma cells in serum-free medium: ethanolamine is an essential component. Proc Natl Acad Sci USA 79(4):1158–1162

Nyberg GB, Balcarcel RR, Follstad BD, Stephanopoulos G, Wang DIC (1999) Metabolic effects on recombinant interferon-χ glycosylation in continuous culture of chinese hamster ovary cells. Biotechnology and Bioengineering 62(3):336–347

Pavlou AK (2004) The market of therapeutic recombinant proteins to 2010. J Commer Biotechnol, 10(4):363

Rasmussen B, Davis R, Thomas J, Reddy P (1998) Isolation, characterization and recombinant protein expression in veggie-CHO: A serum-free CHO host cell line. Cytotechnology 28(1):31

Schlaeger EJ (1996) The protein hydrolysate, Primatone RL, is a cost-effective multiple growth promoter of mammalian cell culture in serum-containing and serum-free media and displays anti-apoptosis properties. J Immunol Meth 194(2):191–199

Sung YH, Lim S, Chung JY, Lee GM (2004) Yeast hydrolysate as a low-cost additive to serum-free medium for the production of human thrombopoietin in suspension cultures of chinese hamster ovary cells. Applied Microbiology and Biotechnology 63(5):527–536

Van Brunt J (1986) Immobilized mammalian cells: The gentle way to productivity. Nat Biotech 4(6):505

Verhoeye F, Burteau C, Mols J, Ballez J-S, Bare G, Thonart P, Bastin G, Agathos SN, Schneider Y-J (2001) Use of a plant peptone-containing serum-free medium for the cultivation of CHO cells in suspension and on microcarriers. In: Linder-Olsson E et al (eds) Animal cell technology: from target to market. Kluwer, Dordrecht, p 362

Warburg O, Negelein E (1924) Ueber den stoffwechsel der tumoren. Biochem Z 153:319–344

Werner RG (2004) Economic aspects of commercial manufacture of biopharmaceuticals. J Biotechnol 113(1–3):171

White P (1946) Cultivation of animal tissues *in vitro* in nutrients of precisely known constitution. Growth 10:231–289

Wurm FM (1997) Aspect of gene transfer and gene amplification in recombinant mammalian cells. In: Hauser H, Wagner R (eds) Mammalian cell biotechnology in protein production. Waltar de Gruyter, Berlin, pp 87–120

Wurm FM (2004) Production of recombinant protein therapeutics in cultivated mammalian cells. Nat Biotechnol 22(11):1393–1398

Zhang Y, Zhou Y, Yu J (1994) Effects of peptone on hybridoma growth and monoclonal antibody formation. Cytotechnology 16 (3):147–150

Chapter 4
Benefits and Limitations of Protein Hydrolysates as Components of Serum-Free Media for Animal Cell Culture Applications

Protein Hydrolysates in Serum Free Media

Juliet Lobo-Alfonso, Paul Price, and David Jayme

Abstract Increased understanding of influential factors for the cultivation of animal cells, combined with heightened regulatory concern over potential transmission of adventitious contaminants associated with serum and other animal-derived components, has elevated interest in using protein hydrolysates as serum replacements or nutrient supplements. This paper reviews the chemistry and biology of various hydrolysates derived from animal, plant and microbial sources. It provides specific examples of a beneficial selection of plant and yeast hydrolysates as ingredients of serum-free nutrient formulations for bioproduction applications of cultured mammalian and insect cells. Strategies for customizing and optimizing nutrients for specialized applications and general benefits and limitations of protein hydrolysates for biopharmaceutical production are also discussed.

Keywords Nutrient medium formulation • Protein-free media • Chemically-defined media • Biochemical characterization • Biological production • Serum replacement • Hydrolysates as supplements

Introduction

In the early days of animal cell cultivation, almost all culture media required animal serum as a growth-promoting component. Serum was included in the medium as a source of nutrients, hormones, growth factors and protease inhibitors. Serum facilitated the attachment and spreading of cells and provided non-specific protection against mechanical damage and shear forces, bound toxic compounds and improved the buffering capacity of the medium. Albumin and transferrin in the serum acted as carriers of lipids, fatty acids, hormones and trace elements (especially iron). Despite these advantages, there were several problems with the use of serum for cell culture,

J. Lobo-Alfonso (✉), P. Price, and D. Jayme
GIBCO Cell Culture, Invitrogen Corporation, Grand Island, NY, USA
e-mail: Juliet.Lobo-Alfonso@lifetech.com

V.K. Pasupuleti and A.L. Demain (eds.), *Protein Hydrolysates in Biotechnology*,
DOI 10.1007/978-1-4020-6674-0_4, © Springer Science+Business Media B.V. 2010

the greatest of which was the regulatory concern due to potential contamination with adventitious agents associated with serum and animal derived products (Merten 1999, 2002; Wessman and Levings 1999). Furthermore, interference with downstream processing and product purification and the inconsistency of raw materials, combined with the limited supply and unpredictable costs of serum, have been driving forces for finding alternatives to serum-supplementation (Murakami 1984).

Over the last 20 years, animal cell culture medium development has progressed towards the replacement or reduction of serum (Barnes and Sato 1980) with simpler, more defined components, such as hormones and attachment factors, insulin, transferrin or metal chelators, lipids, recombinant proteins, and nutrients obtained from synthetic, plant or microbial sources (Jayme 1999; Mizrahi and Lazar 1991). The main objective of serum-free medium development is to replace the undefined serum components with other more defined supplements that provide the essential functions of serum. However, cell growth and metabolism may be significantly different in serum-containing and serum-free media (Chua et al. 1994) with precise nutrient requirements specific for each cell line. Bioproduction and protein quality characteristics may also be affected by transition to serum-free medium (Merten et al. 1999).

Most media used for serum-free culture contain one or several proteins. Serum albumin was often employed as a protein supplement and is still used in many different media. However, other proteins are frequently required for the growth of specific cells, such as insulin and transferrin. Substitutes for these protein supplements include hydrolysates which are peptide fragments, created from hydrolysis of source proteins from plants, yeast, and animal tissue. These serum substitutes which may help to promote growth of many different cell lines have as yet not been clearly defined with respect to their chemical composition, origin and source or identified active factors. They may contain components of animal origin, thus suffering from problems (e.g., lack of biochemical definition and complex regulatory issues) similar to those of serum.

Hydrolysates are obtained by cleaving proteins from the different sources by exogenous proteolytic enzymes such as pepsin or papain; by hydrolysis using strong acids; or by cell autolysis using controlled temperature or osmotic changes, followed by protein digestion by endogenous enzymes. Enzymatic hydrolysis is more gentle than acid hydrolysis and will digest peptide bonds specific to the cleavage site. For example, pepsin will digest at an amide linkage site that involves the aromatic amino acids, phenylalanine, tryptophan, and tyrosine, resulting in a proteolytic digest of a mixture of free amino acids and polypeptides of different lengths. Vitamins, carbohydrates and other micronutrients present in the lysis medium are also included. Acid hydrolysis, which includes a combination of harsh acid and high temperature treatment, results in a mixture where the vitamins are generally destroyed and where some of the more sensitive amino acids such as tryptophan, are also compromised.

The quality of the resultant protein hydrolysate (general term used for *peptones, peptides, hydrolyzed proteins*) is thus defined by the starting raw material, hydrolyzing agent used, process parameters and the extent of hydrolysis. The degree of protein hydrolysis is generally defined by the Amino Nitrogen/Total Nitrogen ratio. Clarification, concentration, pasteurization and drying of the hydrolysate are also factors that need to be controlled to achieve product quality and lot-to-lot consistency. In some instances, the hydrolysates are further purified by ultrafiltration or by chemical means such as acetone precipitation to remove non-proteinaceous components. This additional treatment removes some undefined components and insoluble materials, reduces the need for excessive filtration and minimizes interference with chemical analytical methods, such as endotoxin testing.

Protein hydrolysates (Table 1) have been used with a number of cell lines, both individually and in conjunction with other factors/peptides, to enrich the medium nutritionally, increase the stability of glutamine, and enhance viable cell densities, culture viability and productivity. The very small oligopeptides, low molecular weight substances (Anborski and Moskowitz 1968; Deparis et al. 2003; Yamane and Murakami 1973) and the free amino acids contribute to the nutritional value (Heidemann et al. 2000; Nyberg et al. 1999; Pasupuleti 2000; Schlaeger 1996) and productivity (Franek et al. 2000; Sung et al. 2004) of the medium. These peptides may also mask the presence of toxic levels of inorganic ions by exerting a protective effect. In addition to being a source of peptides and free amino acids, some hydrolysates may act as a source of fatty acids such as oleic (Cohen et al. 1941), arachidonic, linoleic and linolenic acid (Demain et al. 1959), sterols and phospholipids (Iscove and Melchers 1978; Thompson and Okuyama 2000). Besides supporting the growth and nutrition of cells in culture, peptone hydrolysates have been shown to play a role in cell function and physiology. For example they have been shown to stimulate cholecystokinin secretion and gene transcription in the intestinal cell line STC-1 (Cordier-Bussat et al. 1997); suppress cholesterol uptake in Caco-2 cells (Nagaoka et al. 1999); exhibit opioid activity (Yoshikawa et al. 2003);

Table 1 A selection of protein hydrolysates from various sources and their application in cell culture

Source	Reference
Traditional Animal/meat derived	
Primatone RL, HS	Mizrahi 1977; Schlaeger 1996; Velez et al. 1986
Tryptose Phosphate Broth	Reuveny et al. 1980
Peptone/s	Cordier-Bussat et al. 1997; Jan et al. 1994
Lactalbumin Hydrolysate	Reuveny et al. 1982; Schlaeger et al. 1993
Milk products	Fassolitis et al. 1981; Steiner and Klagsbrun 1981
Casein hydrolysates	Nemoz-Gaillard et al. 1998; Rival et al. 2001; Yamada et al. 1991
Invertebrate derived	
Sericin	Terada et al. 2002
Plant Derived	
Vegetable proteins	Mizrahi and Shahar 1977
Soy peptone/Tryptic soy hydrolysates/ soy flour	Borys et al. 2001; Mizrahi 1977; Murakami et al. 1982
Rice hydrolysates	Bare et al. 2001; Verhoeye et al. 2001
Wheat peptides	Burteau et al. 2003; Simpson et al. 2001
Derived from microbes	
Yeast extract	Fardon et al. 1973; Wyss 1979

suppress lipid peroxidation activity in vitro (Kato et al. 1998) and protect cells and proteins from freezing stresses (Tsujimoto et al. 2001).

Several studies (Heidemann et al. 2000; Katsuta and Takaoka 1973, 1977; Taylor and Parshad 1977) have reviewed the utilization of various peptones and extracts of yeast as basal medium supplements (Keay 1976) for cell culture medium development to sustain growth and bioproduction of various cell lines. While Primatone, casein digest, lactalbumin hydrolysate (LAH), tryptose phosphate broth (TPB) and other hydrolysates derived from animal sources have been demonstrated to promote the growth of cultured cells, this chapter will emphasize hydrolysates of yeast and plant origin, due to their reduced regulatory risks for introducing adventitious agents.

Results

Protein hydrolysates of plant origin, such as molasses and soy, have been known to have excellent nutritional characteristics and are a good source of carbohydrate, potassium, calcium, trace minerals, vitamins, amino acids, oligopeptides, iron salts, lipids, nucleosides, nucleotides and trace low molecular weight components. The biochemical profile of the hydrolysates differs depending on the source and methods of hydrolysis and processing (Table 2). For example, the phosphate content of the hydrolysates listed varied from 0.16% to 4.21%, a difference of over 25-fold. Substantial differences in free amino acid content, AN/TN ratio, and ionic composition were also noted. These biochemical differences, reflective of differences in starting materials and processing methods, might be expected to contribute to variable cell culture performance and highlight the importance of evaluating various hydrolysates for the intended applications. Using previously described growth performance assays common to our serum-free medium development efforts, we explored the benefits and limitations of various hydrolysates as serum-replacements.

Insect Cell Culture

Yeast extract has been used extensively not only in the cultivation of insect cells, but also in the health food industry as a source of vitamins, more specifically the B-complex group. Yeast extract has been obtained from either Brewers yeast, a by-product of the brewing industry, or from Bakers yeast as a by-product of the food industry. Unlike the manufacture of most other peptones, yeast extract is obtained by autolysis of the organism induced by either temperature or osmotic shock. The soluble fraction of the disrupted yeast cell is separated from insoluble cell wall components by centrifugation and filtration. This soluble fraction is then spray dried and provided as a nutrient additive for insect cell culture applications.

Table 2 Variation in the biochemistry of hydrolyzed proteins[a]

Product name	Primatone	Primatone SG	N-Z-Amine	Pepticase	Edamin	Hy-Soy	Corn gluten	Cottonseed	Yeast extract
Source	Meat	Gelatin	Casein	Casein	Lactalbumin	Soy	Corn	Cottonseed	Yeast cells
Method of digestion	Enzyme	Enzyme	Enzyme	Enzyme	Enzyme	Enzyme	Enzyme	Enzyme	Autolytic
αAmino Nitrogen (AN)(%)	7.6	2.8	6.5	4	6.9	1.9	2.6	2.9	4
Total Nitrogen (TN)(%)	13.1	16.5	13.1	13.3	12.3	9.5	10.1	8.7	9.1
Ratio (AN/TN)	58	17	49.6	30.1	56.1	20	25.7	33.3	44
Free amino acids (% of total amino acids) (%)	46	28	54	30	54	13	26	42	25
100–200 Da (%)	41.7	13	54.2	29	21.7	25.4	41	37	38
200–500 Da (%)	48	51	36.2	47.7	42.2	57.5	43	41	40
500–1,000 Da (%)	10.4	30	8.8	22.9	27.7	16.8	13	18	17
Average MW	287.3	423	246.7	367.1	384.5	352.6	310	334	325
Ash (%)	6.8	3.1	4.9	8.2	4	9.2	3.4	8.1	11
Moisture (%)	1.5	2.9	3.4	3.2	3.2	2.8	3	3.4	3
pH	6.4	6.6	6.6	7.1	6.2	7.3	6	6.3	5.6
Sodium (%)	2.73	0.22	2.41	3.35	0.95	2.7	10.7	8.5	0.4
Potassium (%)	1.23	0.14	0.09	0.09	0.91	3	0.11	1.29	3.5
Calcium (mg/g)	0.01	0.09	0.02	0.04	0.67	0.2	0.25	0.25	0.2
Magnesium (mg/g)	0.02	0.7	0.02	0.05	0.03	0.25	0.04	0.47	0.4
Chloride (mg/g)	1.5	0.45	1.02	1.1	0.56	2.72	1.03	2.11	0.5
Sulfate (mg/g)	0.19	1.78	0.08	0.04	0.08	0.36	0.11	0.18	0.07
Phosphate (mg/g)	2	0.33	2.27	4.21	1.23	0.74	0.28	0.16	3
Alanine (mg/g)	52.9	83.8	30.1	32.9	37.8	22.9	52.6	24.6	46.2
Arginine (mg/g)	55.1	75.9	31.3	40.1	21.9	50	20.6	68	35.9
Aspartic acid (mg/g)	65.2	58.1	67.4	59.3	82.8	76.2	27.1	61.4	65.9
Cysteine (mg/g)	4.2	0.6	3	3.1	21	6.1	2.7	2.4	2
Glutamic acid (mg/g)	99.3	107	185.8	166.9	161.5	126.5	117.4	112.1	122.1

(continued)

Table 2 (continued)

Product name	Primatone	Primatone SG	N-Z-Amine	Pepticase	Edamin	Hy-Soy	Corn gluten	Cottonseed	Yeast extract
Source	Meat	Gelatin	Casein	Casein	Lactalbumin	Soy	Corn	Cottonseed	Yeast cells
Method of digestion	Enzyme	Enzyme	Enzyme	Enzyme	Enzyme	Enzyme	Enzyme	Enzyme	Autolytic
Glycine (mg/g)	81	204.5	18.9	17.7	17.4	23.7	17.5	27.1	33.1
Histidine (mg/g)	17.4	8	22.2	16.8	13.4	13.9	13	15.7	9.2
Isoleucine (mg/g)	24	16.3	43.6	52.1	42.5	26	26	18.4	30.3
Leucine (mg/g)	49	30.9	74.6	81.1	92.6	42.6	94.8	32.2	46.3
Lysine (mg/g)	58	38.6	67.7	74.5	75	36.7	11.8	26.2	44.5
Methionine (mg/g)	14.2	7.5	27.2	21	21.4	10.4	11.5	8.2	11.1
Phenylalanine (mg/g)	21.5	20.9	39.8	45.4	30.4	25.5	34.2	30	26.4
Proline (mg/g)	87.6	120.6	88.4	81.5	38.6	31.5	57.5	21.2	19.8
Serine (mg/g)	36.4	30.5	51.4	46.2	37	30.8	28.3	25.5	18.1
Threonine (mg/g)	26.5	17.5	41.8	44.4	36.5	25	18.9	19.2	13.7
Tryptophan (mg/g)	3.8	2.6	9.8	11.2	13.5	6.2	3	5.9	11.9
Tyrosine (mg/g)	6.9	8.6	28.2	26	22.2	18.2	29	16.3	28
Valine (mg/g)	31.5	26.6	59	61.8	51.9	25.6	28.4	28.7	39
Total amino acids (mg/g)	734.5	858.5	890.2	882	817.4	597.8	594.3	543.1	603.5

[a]Data selected from Sheffield catalog, Quest International, 1993

We screened various suppliers of such lysates for growth support of relevant insect cell lines. For the initial screen, we examined two different yeast extract vendors and analyzed test samples both for various elements of biochemical composition and for biological performance. Table 3 summarizes select data from these comparative studies.

Multiple analytes were evaluated, similar to properties described in Table 2. Only those analytes exhibiting significantly different biochemical composition are listed in Table 3. As noted, material from the two vendors was vastly different in sodium content, which, given the (w/v) supplementation level of 6 g/L, profoundly impacts overall medium osmolality. Material from one vendor exhibited a 27% elevation in zinc content. Zinc has been shown to be an important metal ion for maintaining growth and forestalling apoptosis in various cell types in serum-free culture, presumably due to its inhibition of the caspase cascade (Ganju and Eastman 2003). Minor but significant variations were also observed in total protein content.

To correlate these biochemical differences with bioperformance, we analyzed proliferation of three model cell types: High Five™ cells, derived from cabbage louper (*Trichoplusia ni*), and Sf9 and Sf21 cells, both derived from army worm (*Spodoptera frugiperda*). Each cell line was cultivated in the appropriate serum-free medium supplemented (0.6% (w/v)) with test hydrolysate for three subcultures prior to quantitative performance evaluation. We observed that the samples from Vendor A yielded differential growth of these target cells from Vendor B. Hydrolysate samples from Vendor B supported superior growth of High Five cells, but vastly inferior support of Sf9 cells and somewhat decreased growth of Sf21 cells.

We found that it was critical to pre-qualify vendors for the growth of various insect cells. Unpublished studies have shown that ultrafiltered yeast as a medium supplement significantly outperforms yeast extract, suggesting that ultrafiltration may remove higher molecular weight materials inhibitory to cell cultivation.

Having demonstrated substantial vendor-to-vendor variation, we now examined multiple lots of nominally identical hydrolysate from the same vendor using our bioperformance assay panel. For this study, we also included D.Mel-2, a fruit fly (*Drosophila melanogaster*) derived cell line, in our bioperformance analysis.

As illustrated in Fig. 1, we reaffirmed our previous observation that different cell types performed differently using the same hydrolysate. However, this assay also revealed that multiple lots from the same vendor also exhibited significant variation. Material from one vendor generally performed well with minimal lot-to-lot variation in support of High Five growth. Other test cell types exhibited substantially greater performance variability. For example, growth of D.Mel-2 cells in serum-free media supplemented with various hydrolysate lots ranged 0–135 in percentage of cell growth relative to control values. Sf9 and Sf21 cells also exhibited broad ranges of performance, although mean values were reduced relative to control performance values.

Table 3 Yeast extract from two different vendors were screened for biochemical and biological performance using High Five™, Sf9, and Sf21 cells

	Biochemistry				Relative proliferative activity of insect cells		
	Sodium	Zinc	Protein	Free fatty acids	% of Control		
	mg/L[a]	mg/L[a]	mg/g	mg/L[a]	High Five™	Sf9 cells	Sf21 cells
Vendor A							
Mean	56.5	11.8	482.3	4.7	72.5	91.5	94.0
±SD	± 13	± 2	± 65	± 0.7	± 9.8	± 14	± 7
Vendor B							
Mean	815.0	15.0	324.3	4.4	113.8	21.8	62.3
±SD	± 26	± 1	± 25	± 0.3	± 12	± 23	± 8

[a]Hydrolysates were prepared as 20% solution and supplemented to the respective yeast extract-free medium to a final level of 6 g/L. The mean represents four lots from each vendor.

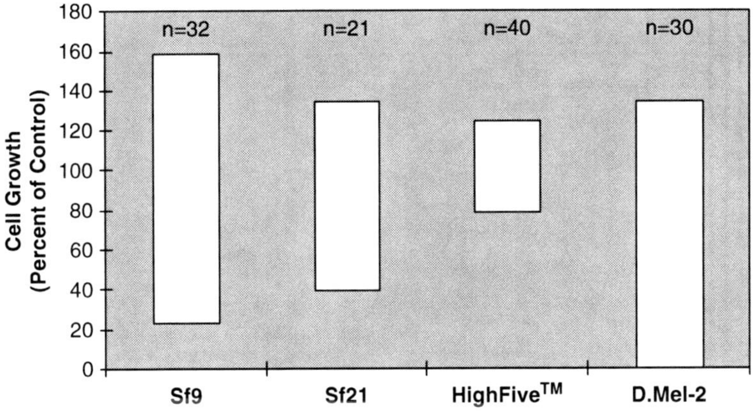

Fig. 1 Prescreening yeast extract raw lots in various insect cell lines (Sf9, Sf21, High Five™ and D.Mel-2). The above graph indicates the minimum and maximum viable cells numbers (expressed as a percent of growth in control medium (0–180%) supplemented with a previously accepted lot of raw material)

Virus Production in Mammalian Cells

Manufacturers of human and veterinary vaccines have become increasingly concerned with biosafety precautions and the necessity of reducing or eliminating risks associated with animal origin materials within the cell culture environment. Historically, vaccine production was performed using relatively simple basal medium formulations supplemented with animal serum and animal origin hydrolysates, such as tryptose phosphate broth (TPB) or lactalbumin hydrolysate (LAH). Current industry trends, however, include migration to serum-free medium containing non-animal-derived hydrolysates.

Studies were therefore initiated to formulate cell culture media devoid of animal-sourced proteins. Enzymatic hydrolysates of a variety of non-animal sources (Table 4) were examined as supplements to the basal medium. Hydrolysates of wheat, soy, rice and extracts from baker's yeast were titrated in the basal media. Cell growth with each hydrolysate was compared to unsupplemented (negative control) or medium supplemented with FBS (positive control). Results were expressed as mean cell count per 25 cm^2 flask over three subcultures and relative growth efficiency (RGE) for each of the medium formulations. RGE was calculated by dividing the mean cell count for a given medium formulation by that of the serum supplemented control.

During our development of a serum-free medium formulation targeted for virus production applications (VP-SFM), we initially screened various plant hydrolysates to determine the preferred candidate to support Vero (African green monkey kidney) cell growth and support production of target viruses. We examined Vero cell growth in a prototype serum-free formulation supplemented with various sources of hydrolysates

Table 4 List of hydrolysates and vendors/products evaluated for development of serum-free media

Source	Hydrolysate name	Vendors
Yeast	Yeast extract	Bectin Dickinson/Difco
	Yeastolate	Champlain
		Global
		Oriental yeast
		Company
		Quest
		Redstar
Soy	Refined soy	DMV
	Hysoy 2NA16	Red Star
	Hysoy Dev. 02997	Quest
	Hysoy UF 1510	Marcor
	Soy 1601	
	Soy Peptone K	
	Soy Peptone II	
Pea	Martone B-1	Marcor
	Veg. Peptone #1	Oxoid
	Refined pea	Quest
		DMV
Rice	5115	Quest
	5603	
Wheat	HyPep 4601 defined wheat	Quest
		DMV
Cottonseed	HyPep 7501	Quest
		Pharmacia
		DMV
Aloe vera		Terra pure

and compared growth following three adaptive passages relative to E-MEM reference medium supplemented with 5% (v/v) FBS. As noted in Fig. 2, several test hydrolysates supported growth equal to or superior to the FBS-supplemented reference materials, while one test hydrolysate (wheat) was readily eliminated by this initial screen.

The poor performance of wheat extract as a medium supplement for animal cells is not surprising. Previous studies have demonstrated that wheat gluten extracts or extracts with high free amino acid content may be toxic or induce toxic effects in certain cell types in vitro (Auricchio et al. 1987; Burteau et al. 2003) and can inhibit protein synthesis in cell-free systems of animal cells (Coleman and Roberts 1982).

Further analysis of multiple hydrolysate sources and nutrient optimization studies resulted in commercial development of the VP-SFM formulation. Figure 3 illustrates comparative yields of three target viruses obtained from Vero cell cultures maintained and infected in either serum-free VP-SFM or serum-supplemented controls. Virtually identical titers of sindbis virus, poliovirus 1 and pseudorabies virus were obtained in the two media.

When we expanded our growth qualification studies to other mammalian cell types relevant to human and veterinary vaccine production, we noted that VP-SFM

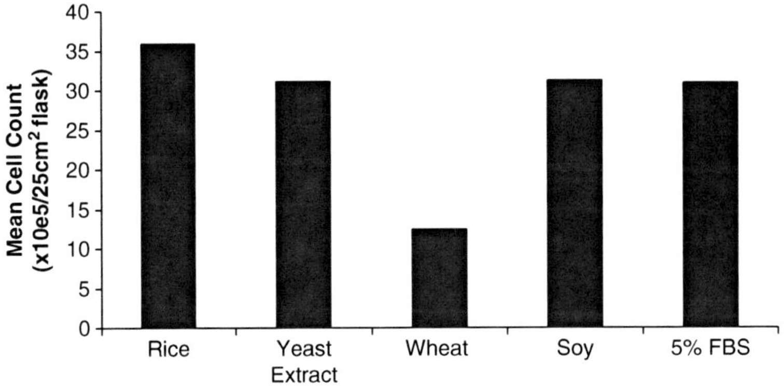

Fig. 2 Comparison of various hydrolysates on growth of VERO cells: Four different hydrolysates were tested in a prototype serum-free medium designed for VERO cells. Each hydrolysate was supplemented into the medium at 200 mg/L

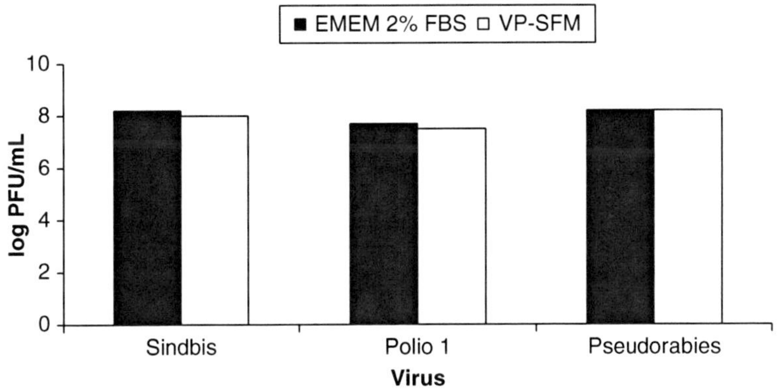

Fig. 3 Virus titration in VERO cells. Production of three model viruses by VERO cells grown in serum-free medium supplemented with a plant hydrolysate was compared with virus production by VERO cells grown in EMEM with 2% serum (control)

performed well in supporting Vero cell culture applications as well as for several other cell types, but that other hydrolysate combinations, together with modifications to the basal formulation, supported cell growth and virus production of a broader range of commercially-relevant kidney epithelial cell types. This observation led to development and commercialization of OptiPRO SFM, a second generation serum-free medium.

The four panels of Fig. 4 illustrate the utility of the hydrolysate-supplemented OptiPRO SFM formulation for cell growth and virus production.

Panel 4a compares growth performance of Vero cells and production of reovirus in OptiPRO SFM and its precursor formulation, VP-SFM. Test results were essentially equivalent for this comparison.

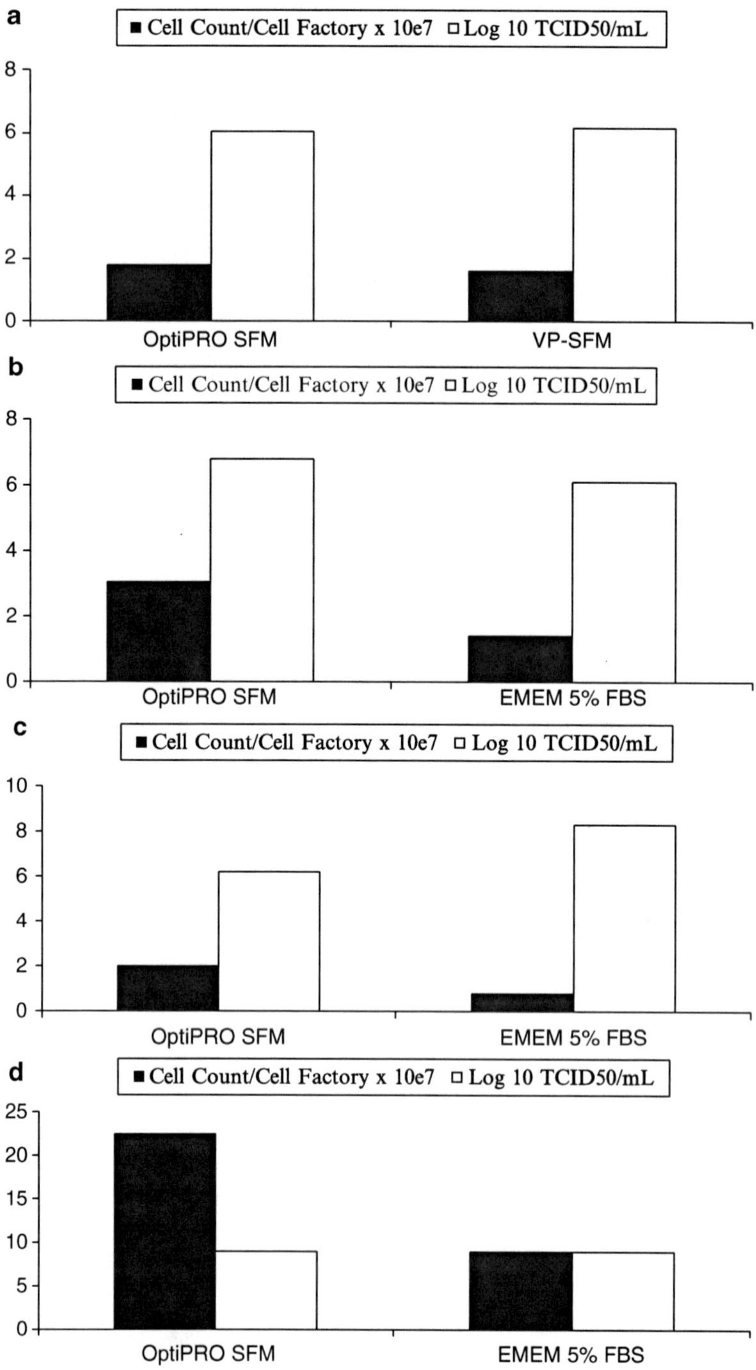
a
Cell Count/Cell Factory x 10e7
Log 10 TCID50/mL
0
2
4
6
8
OptiPRO SFM
VP-SFM
b
Cell Count/Cell Factory x 10e7
Log 10 TCID50/mL
0
2
4
6
8
OptiPRO SFM
EMEM 5% FBS
c
Cell Count/Cell Factory x 10e7
Log 10 TCID50/mL
0
2
4
6
8
10
OptiPRO SFM
EMEM 5% FBS
d
Cell Count/Cell Factory x 10e7
Log 10 TCID50/mL
0
5
10
15
20
25
OptiPRO SFM
EMEM 5% FBS

Panel 4b compares growth of MDCK (Madin-Darby Canine Kidney) cells and production of canine adenovirus in OptiPRO SFM to E-MEM plus 5% FBS. Cell growth in OptiPRO SFM was slightly higher as was total virus production.

Panel 4c compares growth of PK-15 (Porcine Kidney) cells and pseudorabies virus production in OptiPRO SFM and E-MEM plus 5% FBS test media. PK-15, a challenging cell line to cultivate in serum-free medium, exhibited superior growth in OptiPRO SFM and acceptable but reduced virus titers.

Panel 4d compares growth of MDBK (Madin-Darby Bovine Kidney) cells and production of infectious bovine rhinotracheitis (IBR) virus in OptiPRO SFM and E-MEM plus 5% FBS. Cell growth was substantially greater in OptiPRO SFM, due to a surprising loss of contact inhibition. Total virus production was comparable, but the amount of specific virus per cell was reduced in OptiPRO SFM.

Recombinant Protein Production by CHO Cells

To support serum-free production of recombinant proteins in CHO (Chinese hamster ovary) cells in suspension culture, we initially commercialized a serum-free medium named CHO-S-SFM II. However, a critical constituent of this formulation was a bovine glycoprotein that was prohibitively expensive, sourced from animal (bovine) origin materials, and suffered from inadequate supply, and therefore limited its utility for biopharmaceutical production.

Clearly, a primary target of our next generation serum-free medium development program was to eliminate the glycoprotein and replace its functionality with alternative constituents of acceptable cost, availability, and source. Screening of multiple plant-derived hydrolysates led to development of a protein-free formulation, designated CHO III PFM. Concurrent development also permitted elimination of the undefined protein hydrolysate and yielded a protein-free, chemically-defined formulation, designated CD CHO.

Figure 5 illustrates comparative growth of recombinant CHO cells adapted to grow in these two protein-free media. These data indicate that the initial proliferative rate was slightly higher in the hydrolysate supplemented CHO III PFM formulation. Cells grown in the CD CHO formulation ultimately attained comparable peak cell density and sustained cell viability for a longer period, owing to the more nutrient-rich

Fig. 4 (**a**) Growth of cells and virus in OptiPRO SFM (VERO). Cell factories were plated with 8×10^6 VERO cells and counted or inoculated with REO-3 on day 2. The titer of the inoculum was $1 \times 10^8/TCID_{50}/mL$ with a MOI of 0.1 (1 virus/10 cells). (**b**) Growth of cells and virus in OptiPRO SFM (MDCK). Cell factories were plated with 7.2×10^6 MDCK cells and counted or inoculated with CAV on day 2. The titer of the inoculum was $1 \times 10^8/TCID_{50}/mL$ with a MOI of 0.1 (1 virus/10 cells). (**c**) Growth of Cells and virus in OptiPRO SFM (PK-15). Cultures of PK-15 cells inoculated with PRV were carried for four subcultures in respective media, counted and sent for virus titration. Counts represent the average per 2–25 cm^2 flask. The titer of the inoculum was $3.16 \times 10^8/TCID_{50}/mL$ with a MOI of 0.1(1 virus/10 cells). (**d**) Growth of cells and virus in OptiPRO SFM (MDBK). Cell factories plated with 9.6×10^6 MDBK cells and counted or inoculated with IBR on day 4. The titer of the inoculum was $1 \times 10^8/TCID_{50}/mL$ with a MOI of 0.1 (1 virus/10 cells)

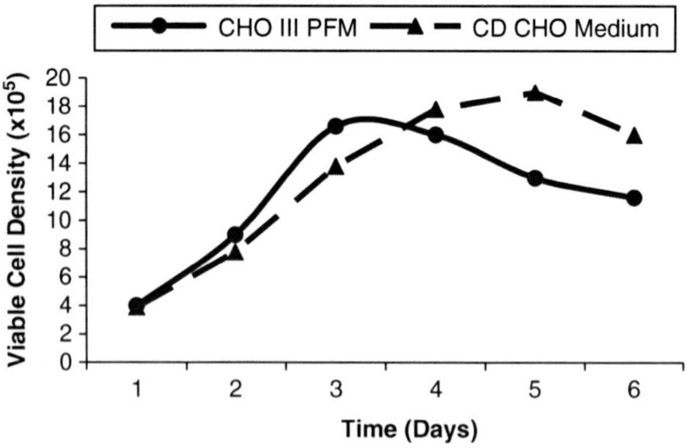

Fig. 5 rCHO cell growth in CHO III PFM versus CD CHO medium. rCHO cells were adapted to grow either in CD CHO Medium, a protein-free, chemically defined medium or in a CHO III prototype, a protein-free medium that contains a plant-derived hydrolysate

environment. Specific protein production rates were comparable in the two media, with volumetric productivity somewhat higher in the CD CHO medium, owing to the extended bioreactor longevity. Not shown is the observation that CHO cells generally exhibited more facile adaptation to the hydrolysate-supplemented CHO III PFM formulation than the hydrolysate-free, chemically-defined CD CHO medium, although CHO cell lines that successfully adapted to CD CHO exhibited excellent growth and biological production characteristics.

Cationic Lipid Transfection

Finally, we investigated the impact of medium composition on an upstream process, specifically, to determine how the presence of protein hydrolysates would affect the ability to transfect CHO-K1 cells using cationic lipid reagents.

In this study, we incubated adherent CHO-K1 cells with DNA and LipofectAMINE for 5 h at 37°C. Transfection efficiency was measured (qualitatively) by histochemical staining for β-galactosidase (Sanes et al. 1986), comparing color generation and cell morphology in test media containing either soybean hydrolysate, human serum albumin, or both (Fig. 6). Transfection efficiency in the control MEM solution (no additions) was approximately 90%. Addition of albumin did not affect transfection efficiency. However, the percentage of transfected cells was reduced to 10% in the presence of soy hydrolysate, with or without supplemental albumin.

Discussion

These results, combined with various other studies, confirm that supplementation of basal nutrient formulations with protein hydrolysates may beneficially support cell proliferation and biological production. However, use of protein hydrolysates

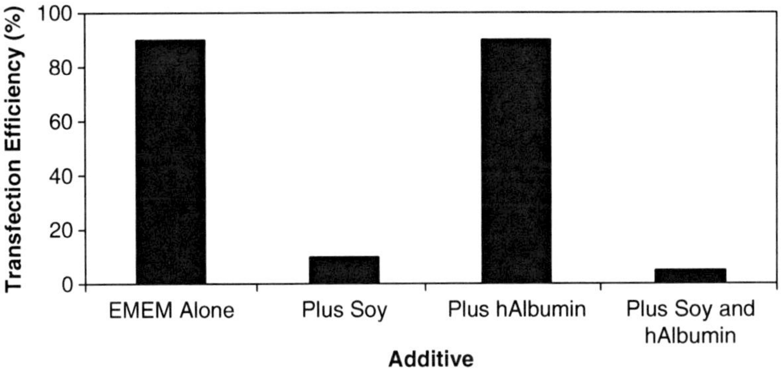

Fig. 6 Transfection efficiency of CHO-K1 cells. The influence of plant hydrolysate (soy) on the cationic lipid transfection efficiency of CHO-K1 cells in the presence and absence of human serum albumin

may be problematic for certain applications. Advantages and disadvantages of protein hydrolysates as cell culture additives are noted below.

Advantages

Technical Benefits

Elimination of the serum additive and optimization of serum-free culture medium for a specialized culture application can be time-consuming and challenging. Serum contributes to cell culture performance by providing growth and attachment factors, supplemental nutrients and buffering capacity, protection against active oxygen species and mechanical shear and various other beneficial factors. Given that the composition of the serum supplement is ill-defined and variable, initial attempts to replace its manifold elements with a chemically-defined cocktail of nutrients may result in only partial success and may demand substantial adaptive weaning of the target cell population.

Creation of a serum-free culture medium by supplementing a basal nutrient formulation with a protein hydrolysate provides an interim solution to this technical challenge. Although still somewhat ill-defined and variable, the hydrolysate provides many of the beneficial factors provided in serum, while reducing some variability and eliminating much of the risk associated with serum-supplemented medium. Hydrolysate addition is a beneficial bridge along the pathway towards development of a fully-defined nutrient environment for the production cell. In addition serum-free media made with protein hydrolysates have been used for the growth of both attachment-dependent and attachment-independent cells (Price and Evege 1997; Price et al. 2002) whereas chemically-defined formulations have primarily been developed for attachment-independent cell growth.

Manufacturing Benefits

Protein hydrolysates are relatively stable and can be stored dry at refrigerated temperatures. As hydrolysates are generally produced at large scale, the large lot size and reasonable consistency of the manufacturing process reduce costs associated with batch qualification of raw materials. Purchasing protein hydrolysates from a reputable, audited supplier may offer some cost advantages compared with the cumulative cost of purified biological components. Careful qualification of hydrolysate raw materials can result in a bioreactor production environment that generates predictably high yields of quality biological product.

Regulatory Benefits

A primary reason for eliminating serum from the cell culture production environment has been the risk of introducing adventitious contaminants. Other factors, such as serum cost and availability, degradation of target molecules by serum proteases, impact on complexity of downstream product purification, potential immunogenicity, and other serum-associated artifacts, may raise concerns. Without question, the primary motivator for biopharmaceutical production processes to eliminate animal serum has been the need to satisfy themselves and regulatory agencies that they have minimized process risks of introducing mycoplasma, virus or prion elements into pharmaceutical product that might adversely impact the human recipient.

While a variety of animal-derived hydrolysates (e.g., lactalbumin hydrolysate, casein hydrolysate, primatone) have historically been incorporated as cell culture nutrient additives for production of vaccines and biologicals, ICH (International Conference on Harmonization) guidelines clearly prefer a culture environment that is free of animal origin materials. Consequently, these milk and meat hydrolysates have been replaced by plant and yeast-derived hydrolysates for most products within the current developmental pipeline to facilitate regulatory approval and avoid the requirement to validate process removal of potential adventitious agents from suspect raw materials.

Disadvantages

Lack of Biochemical Definition

The practical working range for protein hydrolysate supplementation of a basal formulation may range from 100 mg/L to 2.5 g/L, depending upon whether it is used alone or in combination with other nutrient additives. It is evident, therefore, that lot-to-lot variability in biochemical composition or biological performance may have a substantial and variable impact on biomass expansion or product yield.

For an approved biological process, a nutrient formulation that yields consistent bioreactor performance will generally be preferred over an alternative process that may average a higher volumetric productive but exhibits unacceptable variability.

Additionally, the lack of biochemical definition, not unlike its serum precursor, may create undesirable challenges to the downstream purification process. Hydrolysates may complicate protein purification by plugging chromatography columns and by adsorbing non-specifically to binding sites of the column matrix, thus effectively reducing capacity and efficiency. Such complications may be minimized by product-specific capture steps, by pre-filtration, or by implementing higher buffer flow rates or exchange procedures.

Lastly, since regulatory agencies mandate identification and validated process removal of residual materials from the fermentation milieu to minimize adverse host responses, a cell culture nutrient environment that is chemically-defined offers practical benefits to purification and characterization of the harvested biological product.

Processing Complications

As a manufacturer, we have noted that nutrient media supplemented with nutrient hydrolysates typically exhibit extended processing time, due to hydrolysate components with diminished solubility and blinding of sterilizing membrane filters. This extended processing time leads to increased filter usage and potential degradation of sensitive medium components, resulting in adverse impacts on both cost and quality of production.

Production of dry format medium by conventional ball-milling processes is also complicated by inclusion of certain protein hydrolysates, as the elevated temperatures and physical trauma degrade labile hydrolysate constituents. Such limitations may be largely obviated by implementing more novel dry format production processes that reduce residence times, thermal denaturation and mechanical shear, such as hammer milling and fluid bed granulation (Radominski et al. 2001).

Finally, though not a production-scale concern, protein hydrolysates have been reported to interfere with upstream transfection processes with cationic lipids. Investigators working with parental lines adapted to serum-free medium containing protein hydrolysates may wish to evaluate different transfection processes to minimize such complications.

Procurement Concerns

We have experienced disappointment in materials procurement resulting from expending resources to qualify a promising hydrolysate, only to discover that it was obtained from a source material in limited supply or that the manufacturing process to obtain the protein hydrolysate was not economically scalable to meet ultimate production scale requirements. While clearly not a concern for yeast hydrolysates or many plant hydrolysates, we caution routine inquiry regarding manufacturing scale and product availability before committing substantial effort to qualify a more unique hydrolysate.

Regulatory Concerns

Similarly, concerns by regulatory agencies are not fully resolved merely by eliminating serum or animal origin medium components. Biopharmaceutical production requirements mandate assurance that the process vessels used to manufacture the protein hydrolysates have been adequately sanitized and validated treatments employed to eliminate potential cross-contamination from animal origin materials that might be produced using the same processing equipment.

Hydrolysates are not generally produced within a sterile processing environment. Thus, degradation of gram-negative bacteria may introduce lipopolysaccharide constituents, resulting in varying endotoxin levels in the protein hydrolysate. Elevated endotoxin may be problematic from two perspectives. Lipopolysaccharide elements may tend to co-purify with the target biological and may require additional processing and potential yield loss to reduce endotoxin to acceptable dosage levels. Secondly, hydrolysate-supplemented media may interfere with colorimetric assay methods, thus requiring additional evaluative sampling to quantitate artifactual absorbance enhancement or use of the less-sensitive and more cumbersome manual gel clotting method.

Finally, regulatory agencies and biopharmaceutical manufacturers have expanded their animal origin concerns beyond the primary substrates to include processing enzymes. Consequently, to document that a protein hydrolysate is legitimately animal origin free requires demonstration that enzymes used in enzymatic hydrolysis were not derived from animal sources (e.g., trypsin derived from porcine pancreas).

Future Developments

Protein hydrolysates have served as useful cell culture additives for decades. Over the years, there have been significant improvements as source materials and processing methods have evolved with changes in the biotechnology industry. Relatively crude lysates and extracts of animal origin organs, tissues and proteins grand-fathered in historical manufacturing processes are being replaced by more reproducible yeast and plant-derived substitutes as preferred nutrient media additives for biopharmaceutical pipeline products. With the trend towards continuous refinement of raw materials and manufacturing process, we anticipate a persistent requirement for protein hydrolysates to evolve with the industry through incorporation of the following upgrades.

Processing Improvements

The primary manufacturing processes for protein hydrolysates require either enzymatic digestion or autolysis using acid, base or heat. Analogous to the standardization processes (Jayme et al. 1990) that emerged during the 1980s for sourcing, processing and characterization of fetal bovine serum (FBS), there exists an urgent need to assemble an industry consortium to generate guidelines for protein hydrolysates. Although catalyzed by government functions, these FBS standardization panels assembled key industry participants – raw material suppliers, manufacturers and end users – to determine minimal acceptable guidance. While such standards could not totally eliminate inherent batch-to-batch variations in serum, they established foundational definitions and terminologies, acceptable starting materials and manufacturing procedures for aseptic collection and conversion, and assay methods and specifications for high quality material.

To evolve from its current status, the protein hydrolysate industry must convene to establish comparability standards. Anyone who procures hydrolysates as raw materials for biopharmaceutical production should expect that the resultant composition of matter from comparable starting materials and processed by approved manufacturing methods by reputable suppliers should be equivalent. Such bioequivalence among competitive protein hydrolysate suppliers would facilitate the product substitution among approved vendors that dual sourcing, continuity of supply and good manufacturing practices dictate. The excessive variation in biochemical composition and biological performance among nominally comparable protein hydrolysates from different suppliers must be significantly reduced to remain viable raw materials for biopharmaceutical production.

Additionally, to facilitate supplier qualification and audit, there should evolve standard process guidelines. Such standardization should include minimal acceptable guidelines for process sanitization and containment and for quarantine of raw materials and finished goods. It should also evolve to standardize hydrolysis processes and segregation of facility and equipment used to process animal versus non-animal-derived materials.

The definition of "animal origin" continues to evolve from determining whether or not the original primary source material was derived from an animal to encompass secondary exposure, such as nutrient components within the bacterial fermentation broth or animal-derived processing enzymes. The definition must ultimately be embraced by hydrolysate manufacturers, who will be subjected to quality assurance auditors and required to document absence of secondary animal origin materials throughout the manufacturing process.

This compelling trend to eliminate risks of animal-sourced adventitious contaminants from the process environment should ultimately extend both upstream and downstream of the hydrolysis process. Plant-sourced hydrolysates will require assurance of absence of pesticide and herbicide residues, just as current practice requires assurance against exposure to cephalosporin antibiotics. Enhanced manufacturing processes for aseptic handling and containment should also reduce variability in endotoxin content. Finally, as protein additives to cell culture have evolved from organ extracts and blood-derived fractions to recombinant proteins and, ultimately, to biomimetic peptides, we anticipate that increased consistency in hydrolysates will be achieved through improved definition of the source materials.

An anticipated benefit from such standardization will be increased biochemical and bioperformance reproducibility of multiple lots from a given supplier. This increased security will have an associated but acceptable cost: elevating manufacturing process standards to comparability with many other raw materials used for biopharmaceutical production may result in some modest escalation in the price of protein hydrolysates to offset the incremental production cost incurred by suppliers.

Hydrolysate Characterization

Consistent with the quest to improve the degree of biochemical definition of the culture environment, future requirements will likely demand elevated characterization of protein hydrolysates. Historically, analysis of free amino acids provided adequate definition. But with improved characterization methods, e.g., tryptic mapping, mass spectrometry and other analytical instrumentation, more comprehensive qualitative and quantitative analysis of free amino acids, oligopeptides, other nutrients and trace contaminants may be anticipated.

The net result of these superior analytical tools will be protein hydrolysates that are more well-defined, leading to improved lot-to-lot performance consistency. For quality assurance and lot release to manufacturing purposes, a lot-specific certificate of analysis will be required.

Product Development

Product improvement will be a natural consequence of the requirement to migrate from animal origin enzymatic treatments to recombinant proteases for hydrolytic processing. Besides the potential risk of adventitious agents, proteolytic enzymes obtained from animal tissues (e.g., porcine pancreatic trypsin) are notoriously impure and contain contaminating proteases with different peptide cleavage specificities. Use of recombinant proteases to alleviate animal origin risks will facilitate the corollary benefit of providing unique cleavage specificity and more homogeneous hydrolysates.

Possible additional product improvements include larger peptides with superior lipid-carrying potential to substitute for serum albumin or other animal origin proteins historically employed in serum-free formulations for delivery of cholesterol, fatty acids and other bioactive lipids. Alternatively, post-hydrolysis process upgrades may be investigated that might enhance solubility and filterability without adversely impacting biological performance.

Chemically-Defined Medium

Our specialty medium formulation strategy has generally transitioned from an initial serum-supplemented medium to an enhanced formulation that permitted substantial serum reduction, then sequentially to a completely serum-free medium. Taking care to eliminate constituents of animal origin, the prototype serum-free formulation would be further modified to eliminate all protein or polypeptide components to produce a protein-free medium. Such protein-free media would generally contain one or more protein hydrolysates. The ultimate enhancement of these protein-free media would be to eliminate the hydrolysates and generate a nutrient formulation containing only biochemically-defined, low molecular weight ingredients that exhibit comparable stability and biological performance as its precursors.

To accomplish this final phase, reduction in the level of hydrolysate supplementation to basal formulation frequently facilitates greater understanding of cellular nutritional and attachment requirements and accelerates development of the more desirable chemically-defined formulation. This medium optimization process may be beneficially implemented to reduce or eliminate requirement for serum or hydrolysate for challenging cell species. Such defined environments ultimately result in more consistent biomass expansion, specific and volumetric product yields, and bioreactor campaigns. Elimination of undefined components also facilitates greater understanding of the factors associated with improved culture viability and productivity and those associated with regulation of cellular metabolic function, differentiation and apoptosis.

In conclusion, protein hydrolysates may be beneficially supplemented to nutrient medium to enhance animal cell proliferation and biological production. Hydrolysates derived from plant and yeast sources largely overcome many regulatory concerns regarding introduction of adventitious agents into the biological production environment. Prospective users must be aware of sourcing and processing differences between vendors and screen for lot-to-lot variations in biochemical composition and biological performance to ensure suitability for their specific application, as recommended for any ill-defined additive. Standardization of hydrolysate processing and analytical testing methods will improve suitability for biopharmaceutical production. While improvements in development of chemically-defined media and metabolic engineering of production cell lines may eventually obviate the necessity of hydrolysates, they currently serve a highly useful role in improving biomass expansion and product yield.

Acknowledgments The authors wish to thank Steve Gorfien, Mary Lynn Tilkins, Douglas Danner and Philip Grefrath for data used in this review.

References

Anborski RL, Moskowitz M (1968) The effects of low molecular weight materials derived from animal tissues on the growth of animal cells in vitro. Exp Cell Res 53:117–128

Auricchio S, De Ritis G, Vincenzi M De, Latte F, Maiuri L, Pino A, Raia V, Silano V (1987) Prevention by mannan and other sugars of in vitro damage of rat fetal small intestine induced by cereal prolamin peptides toxic for human celiac intestine. Pediatr Res 22:703–707

Bare G, Charlier H, De Nijs L, Verhoeye F, Schneider YJ, Agathos S, Thonart P (2001) Effects of rice protein hydrolysate on growth of CHO cells and production of Human Interferon-γ in a serum-free medium. In: Lindner-Olsson E, Chatzissavidou N, Luellau E (eds) Animal cell technology: from target to market. Kluwer, Netherlands, pp 217–219

Barnes D, Sato G (1980) Methods for growth of cultured cells in serum-free medium. Anal Biochem 102:255–270

Borys MC, Hughes KD, Ryan JM (2001) The effects of different plant protein hydrolysates on SP2/O cells expressing recombinant Pro-urokinase. In Vitro Cell Dev Biol Anim 37(3, Part II):VT-1000

Burteau CC, Verhoeye FR, Mols JF, Ballez J-S, Agathos SN, Schneider YJ (2003) Fortification of a protein-free cell culture medium with plant peptones improves cultivation and productivity of an interferon -γ-producing CHO cell line. In Vitro Cell Dev Biol Anim 39:291–296

Chua F, Oh SKW, Yap M, Teo WK (1994) Enhanced IgG production in eRDF media with and without serum: A comparative study. Methods 167:109–119

Cohen S, Snyder JC, Mueller JH (1941) Factors concerned in the growth of corynrbacterium diphtheriae from minute inocula. J Bacteriol 41:581–591

Coleman WH, Roberts WK (1982) Inhibitors of animal cell-free protein synthesis from grains. Biochim Biophys Acta 696:239–244

Cordier-Bussat M, Bernard C, Haouche S, Roche C, Abello J, Chayvialle J-A, Cuber JC (1997) Peptones stimulate cholescystokinin secretion and gene transcription in the intestinal cell line STC-1. Endocrinology 138:1137–1144

Demain AL, Hendlin D, Newkirk J (1959) Role of fatty acids in the growth stimulation of Sarcina species by vitamin-free casein digest. J Bacteriol 78:839–843

Deparis V, Durrieu C, Schweizer M, Marc I, Goergen JL, Chevalot I, Marc A (2003) Promoting effect of rapeseed proteins and peptides on Sf9 insect cell growth. Cytotechnology 42(2):75–85

Fardon JC, Poydock SME, Tsuchiya Y (1973) The effect of a yeast extract (PCO) on the mitotic activity of neoplastic and normal cells in vitro. J Surg Oncol 5:307–314

Fassolitis AC, Larkin EP, Novelli RM (1981) Serum substitute in epithelial cell culture media: Nonfat dry milk filtrate. Appl Environ Microbiol 42:200–203

Franek F, Hohenwarter O, Katinger H (2000) Plant protein hydrolysates: preparation of defined peptide fractions promoting growth and production in animal cell cultures. Biotechnol Prog 16:688–692

Ganju N, Eastman A (2003) Zinc inhibits Bax and Bak activation and cytochrome c release induced by chemical inducers of apoptosis but not by death-receptor-initiated pathways. Cell Death Diff 10(6):652–661

Heidemann R, Zhang C, Qi H, Rule JL, Rozales C, Park S, Chuppa S, Ray M, Michaels J, Konstantinov K, Naveh D (2000) The use of peptones as medium additives for the production of a recombinant Therapeutic protein in High-Density cultures of mammalian cells. Cytotechnology 32:157–167

Iscove NN, Melchers F (1978) Complete replacement of serum by albumin, transferrin, and soybean lipid in cultures of lipopolysacchride reactive ß Lymphocytes. J Exp Med 147:923–933

Jan DC-H, Jones SJ, Emery AN, Al-Rubeai M (1994) Peptone, a low cost growth-promoting nutrient for intensive animal cell culture. Cytotechnology 16:17–26

Jayme DW (1999) An animal origin perspective of common constituents of serum-free medium formulations. In: Brown F, Cartwright T, Horaud F, Speiser JM (eds) Animal sera, animal sera derivatives and substitutes used in the manufacture of pharmaceuticals: viral safety and regulatory aspects, vol 99, Dev Biol Stand. Karger, Switzerland, pp 181–187

Jayme D, Tribby I, Spendlove R, Peterson W (1990) Fetal bovine serum: proposed guideline. National Committee for Clinical Laboratory Standards: Subcommittee Report, vol 10, pp 1–38

Kato N, Sato S, Yamanaka A, Yamada H, Fuwa N, Nomura M (1998) Silk protein, sericin, inhibits lipid peroxidation and tyrosinase activity. Biosci Biotechnol Biochem 62(1):145–147

Katsuta H, Takaoka T (1973) Cultivation of cells in protein and lipid-free synthetic media. Meth Cell Biol 6:1–42

Katsuta H, Takaoka T (1977) Improved synthetic media suitable for tissue culture of various mammalian cells. Meth Cell Biol 14:145–158

Keay L (1976) Autoclavable low cost serum-free cell culture media: the growth of established cell lines and production of viruses. Biotechnol Bioeng 18(3):363–382

Merten OW (1999) Safety issues of animal products used in serum-free medium. In: Brown F, Cartwright T, Horaud F, Speiser JM (eds) Animal sera, animal sera derivatives and substitutes used in the manufacture of pharmaceuticals: viral safety and regulatory aspects, vol 99, Dev Biol Stand. Karger, Switzerland, pp 167–180

Merten OW (2002) Virus contamination of cell cultures – a biotechnological view. Cytotechnology 39:91–116

Merten OW, Kallel H, Manuguerra JC, Tardy-Panit M, Crainic R, Delpeyroux F, Van der Werf S, Perrin P (1999) The new medium MDSS2N, free of any animal protein supports cell growth and production of various viruses. Cytotechnology 30(1–3):191–201

Mizrahi A (1977) Primatone R.L in mammalian cell culture media. Biotech Bioeng 19:1557–1561

Mizrahi A, Lazar A (1991) Media for cultivation of animal cells: an overview. In: Sasaki R, Ikura K (eds) Animal cell culture production of biologicals. Kluwer, Netherlands, pp 159–180

Mizrahi A, Shahar A (1977) Partial replacement of serum by vegetable proteins in BHK culture medium. J Biol Stand 5:327–332

Murakami H (1984) Serum-free cultivation of plasmacytoies & hybridomas. In: Barnes DW, Sirbasku DA, Sato GH (eds) Methods for serum-free culture of neuronal & Lymphoid cells. Alan R Liss, New York, pp 197–206

Murakami H, Masui H, Sato GH (1982) Suspension culture of hybridoma cells in serum-free medium: Soybean lipids as essential components. In: Cold Spring Harbor conferences on cell proliferation, growth of cells in hormonally defined media, vol 9, pp 711–715

Nagaoka S, Miwa K, Eto M, Kuzuya Y, Hori G, Yamamoto K (1999) Soy protein peptic hydrolysate with bound phospholipids decreases micellar solubility and cholesterol absorption in rats and Caco-2 cells. J Nutr 129:1725–1730

Nemoz-Gaillard E, Bernard C, Abello J, Cordier-Bussat M, Chayvialle J-A, Cuber J-C (1998) Regulation of cholecystokinin secretion by peptones and peptidomimetic antibiotics in STC-1 cells. Endocrinology 139:932–938

Nyberg GB, Balcarcel R, Follstad BD, Stephanopoulos G, Wang DIC (1999) Metabolism of peptide amino acids by Chinese hamster ovary cells grown in a complex medium. Biotechnol Bioeng 62(3):324–335

Pasupuleti VK (2000) Influence of protein hydrolysates on the growth of hybridomas and the production of monoclonal antibodies. Presented at The Waterside Conference, Miami, FL

Price PJ, Evege EK (1997) Serum-free medium without animal components for virus production. Focus 19:67–69

Price P, Evege E, Nestler L, Grefrath P, Naumovic B, Fatunmbi F, Jayme D (2002) A versatile serum-free medium for kidney epithelial cell growth and virus production. Focus 24:24–28

Radominski R, Hassett R, Dadey B, Fike R, Cady D, Jayme D (2001) Production-scale qualification of a novel cell culture medium format. BioPharm 14(7):34–39

Reuveny S, Bino T, Rosenberg H, Traub A, Mizrahi A (1980) Pilot plant production of human lymphoblastoid interferon. Dev Biol Stand 46:281–288

Reuveny S, Lazar A, Minai M, Feinstein S, Grosfeld H, Traub A, Mizrahi A (1982) Large-scale production of human (Namalva) interferon. Ann Virol 133E:191–199

Rival SG, Fornaroli S, Boeriu CG, Wichers HJ (2001) Caseins and casein hydrolysates. 1. Lipoxygenase inhibitory properties. J Agric Food Chem 49(1):287–294
Sanes JA, Rubenstein JLR, Nicolas J-F (1986) Use of recombinant retrovirus to study post-implantation cell lineage in mouse embryos. EMBO J 5:3133–3142
Schlaeger EJ (1996) The protein hydrolysate, Primatone RL, is a cost-effective multiple growth promotor of mammalian cell culture in serum-containing and serum-free media and displays anti-apototic properties. J Immunol Methods 194:191–199
Schlaeger EJ, Foggetta M, Vonach JM, Christensen K (1993) SF-1, a low-cost culture medium for the production of recombinant proteins in baculovirus infected insect cells. Biotechnol Tech 7:183–188
Simpson NH, Wegkamp HBA, Bulthuis BA, Siemensma AD, Martens DE (2001) Metabolic shifts in hybridoma cells utilizing wheat peptides. In: Lindner-Olsson E, Chatzissavidou N, Luellau E (eds) Animal cell technology: from target to market. Kluwer, Dordrecht, pp 183–184
Steiner KS, Klagsbrun M (1981) Serum-free growth of normal and transformed fibroblasts in milk: Differential requirements of fibronectin. J Cell Biol 88:294–300
Sung YH, Lim SW, Chung JY, Lee GM (2004) Yeast hydrolysate as a low-cost additive to serum-free medium for the production of human thrombopoietin in suspension cultures of Chinese hamster ovary cells. Appl Microbiol Biotechnol 63(5):527–536
Taylor WG, Parshad R (1977) Peptones as serum substitutes for mammalian cells in culture. Meth Cell Biol 15:421–434
Terada S, Nishimura T, Sasaki M, Yamada H, Miki M (2002) Sericin, a protein derived from silkworms, accelerates the proliferation of several mammalian cell lines including hybridoma. Cytotechnology 40:3–12
Thompson GA Jr, Okuyama H (2000) Lipid –linked proteins of plants. Prog Lipid Res 39:19–39
Tsujimoto K, Takagi H, Takahashi M, Yamada H, Nakamori S (2001) Cryoprotective effect of the serine-rich repetitive sequence in silk protein sericin. J Biochem 129(6):979–986
Velez D, Reuveny S, Miller L, Macmillan JD (1986) Kinetics of antibody production in low serum growth medium. J Immunol Methods 86:45–52
Verhoeye F, Burteau C, Mols J, Ballez J-S, Bare G, Thonart P, Bastin G, Charlier H, Agathos S, Schneider YJ (2001) Use of plant peptone-containing serum-free media for the cultivation of CHO cells in suspension and on microcarriers. In: Lindner-Olsson E, Chatzissavidou N, Luellau E (eds) Animal cell technology: from target to market. Kluwer, Dordrecht, pp 362–364
Wessman SJ, Levings RL (1999) Benefits and risks due to animal serum used in cell culture production. In: Brown F, Cartwright T, Horaud F, Speiser JM (eds) Animal sera, animal sera derivatives and substitutes used in the manufacture of pharmaceuticals: viral safety and regulatory aspects, vol 99, Dev Biol Stand. Karger, Switzerland, pp 3–8
Wyss C (1979) Cloning of Drosophila cells: effect of vitamins and yeast extract components. Somat Cell Genet 5(1):23–28
Yamada K, Nakajima H, Ikeda I, Shirahata S, Enomoto A, Kaminogawa S, Murakami H (1991) Stimulation of proliferation and immunoglobulin production by various types of caseins. In: Sasaki R, Ikura K (eds) Animal cell culture production of biologicals. Kluwer, Netherlands, pp 267–274
Yamane I, Murakami O (1973) 6, 8-Dihydroxypurine: a novel growth factor for mammalian cells in vitro, isolated from a commercial peptone. J Cell Physiol 81:281–284
Yoshikawa M, Takahashi M, Yang S (2003) Delta opioid peptides derived from plant proteins. Curr Pharm Des 9(16):1325–1330

Chapter 5
Oligopeptides as External Molecular Signals Affecting Growth and Death in Animal Cell Cultures

František Franek

Abstract Protein hydrolysates in the form of oligopeptides and free amino acids are widely used in animal cell culture for the production of therapeutic proteins. The primary function of protein hydrolysates is to provide nitrogen source and at the same they may increase cell density and higher yields of proteins. It is interesting to note that some peptides exclusively increase cell density, others improve both cell density and product yield, and some peptides suppress cell growth and enhance the product yield. Thus it is very clear that oligopeptides act as external molecular signals affecting growth and death. However, the effect of peptide size and amino acid composition in the protein hydrolysates and the exact mechanism as how this is achieved is still not elucidated in animal cells. In this chapter we describe our work on the fractionation of protein hydrolysates and the use of synthetic peptides on hybridomas. This research work shed some insight about the peptide size, amino acids, concentration and composition of peptides, feeding strategies for peptides but by any means this is not complete and more work needs to be done. For example it is essential to extend this type of work with peptides larger than tetra and penta peptides and with different cell lines to elucidate the mode of action of peptides.

Keywords Oligopeptides • Molecular signals • Apoptosis • Hybridoma • Monoclonal antibody • Protein hydrolysates • Amino acids • Di and tri peptides

Introduction

Production of therapeutic proteins in animal cell bioreactors is increasing, Another expanding field is the production of recombinant viruses for gene therapy. In the development of majority of the mentioned products genetic manipulations of both

F. Franek (✉)
Laboratory of Growth Regulators, Institute of Experimental Botany, Radiová 1, CZ-10227 Prague 10, Czech Republic
e-mail: franek@biomed.cas.cz

V.K. Pasupuleti and A.L. Demain (eds.), *Protein Hydrolysates in Biotechnology*,
DOI 10.1007/978-1-4020-6674-0_5,

the produced molecules (e.g., humanized monoclonals) and the producer cell line (e.g., metabolic engineering of the cell line) play a significant role. However, the availability and the cost of the desired product is dependent not only on the sophisticated modifications of the cell line genome, but also on the volumetric productivity achievable in the bioreactor technology. The concentration of the product at harvesting time largely depends on the set-up of the metabolism that controls the channeling of the nutrients either to biomass synthesis or to product synthesis. Cell productivity is markedly enhanced when cells are prevented from cycling. Thus, apart from genetic modifications of the cell line, manipulation with the cell proliferation rate through providing external molecular signals constitutes a means how to increase the yields of produced proteins.

Crude protein hydrolysates, i.e. mixtures of peptides and of ill-defined contaminating substances, have been mostly considered as inexpensive sources of amino acids improving substantially the nutrition of cultured cells (Mizrahi 1977; Jan et al. 1994; Schlaeger 1996; Nyberg et al. 1999; Heidemann et al. 2000; Bare et al. 2001; Holdread and Brooks 2004).

In our recent work we analyzed the effects of fractions of soy glycinine and wheat gluten hydrolysates obtained by size-exclusion chromatography on Biogel P-2. We have found substantial differences in the activities among individual chromatography fractions of the hydrolysates. The most active fraction was found to be constituted of peptides formed by two to ten amino acid residues.

Thus, we arrived at a conclusion that the hydrolysates did not serve only as a source of utilizable amino acids, but that they also provided peptides exerting specific effects on cell growth and productivity (Franek et al. 2000).

Another argument for the specific-signal character of hydrolysate components emerged from the analysis of positive effects of wheat gluten enzymic hydrolysate fractions on hybridoma cultures. Significant improvement of culture parameters, such as long-term viability and product yield, could be achieved with hydrolysate fractions the amino acid composition of which was rather unbalanced. The sum of dominant amino acids, i.e. glutamic acid/glutamine, leucine and methionine represented about 50 mol%, and most of those amino acids that were known to be essential, from the nutritional point of view, were strongly under-represented (Franek 2004). Parameters of cultures supplemented with various chromatography fractions are given in Table 1.

Table 1 Effects of chromatography fractions of Triticum gluten hydrolysate on hybridoma cultures (Franek et al. 2000; Franek 2004)

Supplement	Viable cells × 10^{-3} cells mL^{-1}	Culture viability (%)	Monoclonal antibody (mg L^{-1})
None (control)	1,180	52	33
Fraction a1	1,390	60	56
Fraction a 21	1,950	64	67
Fraction a 22	1,400	57	58
Fraction a 211	1,840	75	56
Fraction a 212	1,420	73	48

Values on day 6 of the cultures are given. The concentration of peptide fractions at inoculation was 0.2%

The standard way for continuation of this research would be a fine fractionation of the hydrolysates and identification of the most active peptide(s). We chose an alternative and novel approach: instead of laborious and problematic isolation of individual peptides from extremely complex hydrolysates, we started to investigate activities of available synthetic di- to pentapeptides.

Pure synthetic peptides were obtained from Bachem (Bubendorf, Switzerland) or from PolyPeptide Laboratories (Prague, Czech Republic). The activity of individual peptide supplements was tested on the model mouse hybridoma ME-750 producing an IgG2a antibody. The protein-free culture medium was DMEM/F12/RPMI 1640 (3:1:1) supplemented with Basal Medium Eagle (BME) amino acids, 2.0 mM glutamine, 0.4 mM HEPES, and 2.0 g L^{-1} sodium bicarbonate (Franek et al. 1992). The medium was also supplemented with survival-promoting amino acids (Franek and Šrámková 1996a), and with the iron-rich growth-promoting mixture containing 0.4 mM ferric citrate (Franek et al. 1992). In fed-batch cultures, a volume of 0.25 mL of a feeding mixture (DMEM fortified with 10× BME amino acids, 10× BME vitamins, and 20 mM glutamine) was added daily.

The cultures in 25 cm^2 T-flasks were kept at 37°C in a humidified atmosphere with 5% CO_2. The culture volume was 6.0 mL. Monoclonal antibody concentrations were determined by immunoturbidimetry (Fenge et al. 1991) using a calibration curve. The number of apoptotic cells in the cultures was determined by microscopic counting of cells displaying apoptotic morphology, i.e. shrunken cells with ruffled membranes. No swollen necrotic cells were seen in the static cultures used for the assays. The proportion of cell-cycle phases was determined upon permeabilization and staining with propidium iodide.

Results and Discussion

Basic Features of the Action of Synthetic Oligopeptides

For the basic exploration of the effects of pure defined peptides on the course of hybridoma cultures, and, consequently, on the yield of the product, we selected peptides, the composition and molecular mass of which were likely to be similar to those of peptides occurring in protein hydrolysates.

Application of various oligopeptides to the model hybridoma culture showed a marked growth-promoting effect of triglycine and tetraglycine (Table 2). With tetraglycine, we investigated the concentration range at which the peptides exerted their effects. Enhanced viable cell density and higher viability were significant at 4.0 mM tetraglycine. This molar concentration is by several orders of magnitude higher than is the range of active molar concentrations of known peptide hormones. The millimolar range of concentrations at which peptides are able to act on cell culture parameters thus represents a characteristic feature of the action of small oligopeptides on cultured animal cells.

A certain degree of oligomerization is obviously necessary for alanine, in order to enable this amino acid to influence cell growth and production. Trialanine and

Table 2 Effects of pure synthetic peptides on hybridoma cultures

Supplement	Viable cells × 10^{-3} cells mL^{-1}	Culture viability (%)	Monoclonal antibody (mg L^{-1})
None (Control)	1,060	52	30
Concentration dependence of growth-promoting activity (Franek and Katinger 2001; Franek and Katinger 2002)			
Tetraglycine 0.1%	1,640	58	31
Tetraglycine 0.2%	1,980	60	29
Tetraglycine 0.3%	2,090	63	32
Oligomers of a single amino acid (Franek and Katinger 2002)			
Gly	1,080	50	29
Gly-Gly	1,190	52	31
Gly-Gly-Gly	1,490	57	30
Gly-Gly-Gly-Gly	1,720	58	29
Ala	1,030	54	31
Ala-Ala	1,160	70	38
Ala-Ala-Ala	1,450	78	43
Ala-Ala-Ala-Ala	1,480	75	48
Peptides of non-proteinogenic amino acids			
D-Ala-D-Ala-D-Ala	1,510	77	45
β-Ala-β-Ala-β-Ala	1,390	72	38
Peptides enhancing cell density and product yield (Franek and Katinger 2001; Franek and Katinger 2002)			
Ser-Ser-Ser	1,470	69	41
Thr-Thr-Thr	1,190	66	41
Val-Val-Val-Val	1,200	72	39
Pro-Gly-Gln-Gly-Gln	1,260	68	51
Peptides selectively enhancing cell density (Franek et al. 2003)			
Gly-Phe-Gly	1,350	65	29
Gly-Glu-Gly	1,270	81	30
Peptides selectively enhancing product yield (Franek et al. 2003)			
Gly-Lys-Gly	930	62	49
Lys-Lys-Lys	1,130	56	39
Gly-His-Lys	1,110	54	38
Gly-Gly-Lys-Ala-Ala	1,120	75	44

Values on day 6 of batch cultures are given. If not specifically indicated, the concentration of peptides at inoculation was 0.2%

tetraalanine enhanced significantly both the viable cell density and the monoclonal antibody yield (Table 2). The stimulating activity of L-alanine oligomers can by no means be interpreted in terms of improved nutrition, because alanine is a non-essential amino acid that is known to be produced, as a catabolic product, by cultured animal cells (Franek and Šrámková 1996b; Hiller et al. 1993). The finding of growth-stimulating activity of oligomers of non-proteinogenic amino acids D-alanine and β-alanine (Table 2) provides another argument against the view that the positive effects of peptides are based on their providing L-amino acids needed as building stones for synthesis of new proteins.

Investigation of the effects of oligopeptides composed of various amino acids demonstrated the relative independence of the two activities: stimulation of cell growth resulting in higher biomass production, and stimulation of secreted protein production. While some peptides enhanced both the viable cell density of the culture and the final antibody yield (oligoalanines, triserine, trithreonine, glutamine-containing pentapeptide), selective enhancement of cell density was observed with oligoglycines, and with the peptides Gly-Phe-Gly and Gly-Glu-Gly. Addition of peptides which did not stimulate cell growth at all, such as the peptides containing the lysine residue, resulted in high yields of monoclonal antibody (Table 2).

Suppression of Programmed Cell Death

The data in Table 2 show that in the presence of any of the tested peptides, the value of cell viability at the end of the assay period is higher then the value in the control culture.

At the decline phase of batch cultures animal cells tend to actively reduce the viable population size by setting into action the process of programmed cell death. This process, called apoptosis, allows to prevent a sudden collapse of the culture, that would inevitably follow the total exhaustion of nutrients. The suicide mechanism of apoptosis allows the lymphoid cell line to survive as a smaller population under conditions of starvation (Franek et al. 1992; Franek and Dolníková 1991; Vomastek and Franek 1993; Mercille and Massie 1994; Singh et al. 1994; Franek 2003). Peptide supplements thus also act as agents postponing the apoptotic death of the cultures.

Feeding the Peptide-Supplemented Cultures

Supplementation of the cultures with a peptide and feeding with a medium concentrate were found to act synergistically to increase the product yield. By feeding, the antibody yield could be raised from 35 mg L^{-1} in the un-fed control to 60 mg L^{-1} in the fed tetraalanine-supplemented culture (Fig. 1).

Similarly, the antibody yield of 29 mg L^{-1} in the un-fed control increased up to 163 mg L^{-1} in a prolonged fed Gly-Lys-Gly supplemented culture (Franek and Katinger 2002) (Fig. 2).

Tetraglycine itself did not enhance the basic monoclonal antibody yield (51 mg L^{-1}) in batch culture. However, in a culture run in fed-batch mode the final value of monoclonal antibody was 148 mg L^{-1} (Franek and Katinger 2002).

In contrast to the effects of above reported peptides, monoclonal antibody production in the presence of the peptide Gly-Phe-Gly could not be improved by feeding, even though the growth-promoting effect of this peptide was rather pronounced (Franek et al. 2003).

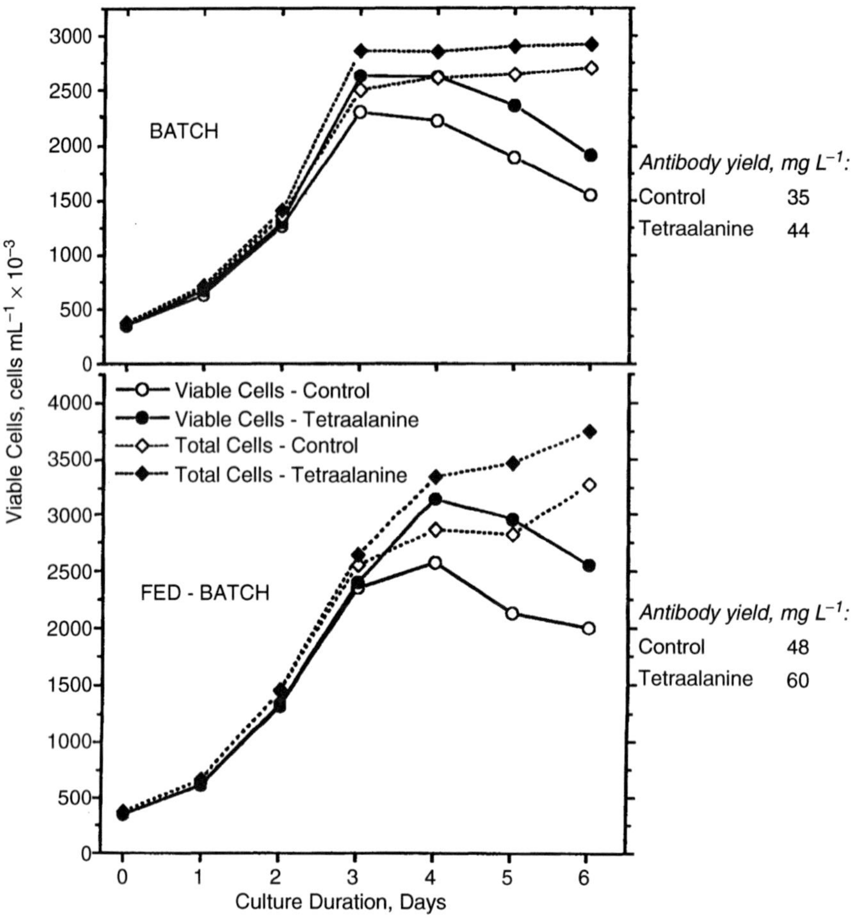

Fig. 1 Batch culture and fed-batch culture in the presence of 0.1% tetraalanine

Intact Peptide Cannot Be Replaced by Amino Acids Constituting the Peptide

Cultures were set up, in which single doses of either the tripeptide Gly-Lys-Gly or of the constituting amino acids glycine and lysine in free form were applied upon culture inoculation. While the viable cell density in the presence of the peptide was lower than in the control, cell density exceeding the control was reached upon addition of the glycine/lysine mixture. With the glycine/lysine mixture a substantially lower monoclonal antibody yield was obtained than that reached with the peptide (Fig. 2) (Franek et al. 2003).

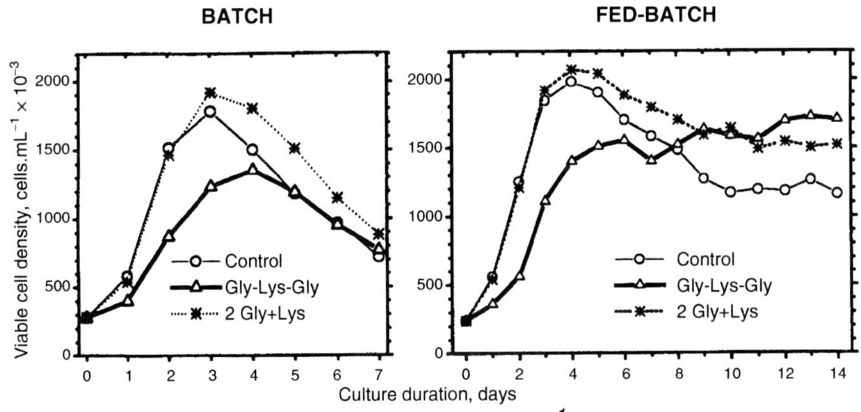

Final MAb concentrations, mg L^{-1}:

	Batch		Fed-batch
Control	29	Control	85
Gly-Lys-Gly	47	Gly-Lys-Gly	163
2 Gly+Lys	34	2 Gly+Lys	98

Fig. 2 Batch culture and fed-batch culture in the presence of 0.2% of the peptide Gly-Lys-Gly, or of an equivalent concentration of free amino acids glycine and lysine

Integrity of the Peptides in the Culture

Tetraglycine and tetraalanine concentrations were found to decrease slightly during a 4 day culture period: tetraglycine to 92%, and tetraalanine to 70% of the starting values, respectively. In parallel, the concentrations of monomeric glycine or alanine increased. We assume that fractions of the tetrapeptides were cleaved by peptidases into amino acids. The balance of alanine was near zero, the balance of glycine indicated a certain consumption (Franek and Katinger 2002).

During a 4 day period, the concentration of the peptide Gly-Lys-Gly decreased by 63%. The concentration of free amino acids, constituting the peptide, increased in comparison with the levels on day 0. New peaks of peptides, most likely dipeptides, emerged in the analyzer records of the supernatants (Franek et al. 2003). Although the lysine-containing peptides were found to be more accessible to proteolytic enzymes than tetraglycine or tetraalanine, the impact of a single dose could be observed during 6 or 7 days of the culture. A hypothesis may be raised that a single peptide dose, if large enough, may provoke long-lasting shifts in the metabolism, including possible alterations in gene expression.

Shift in the Cell-Cycle Phases Distribution

The cell-cycle-phase profiles obtained on day 3, i.e. at the end of the exponential phase, were characterized by an increase of the S-phase fraction in cultures

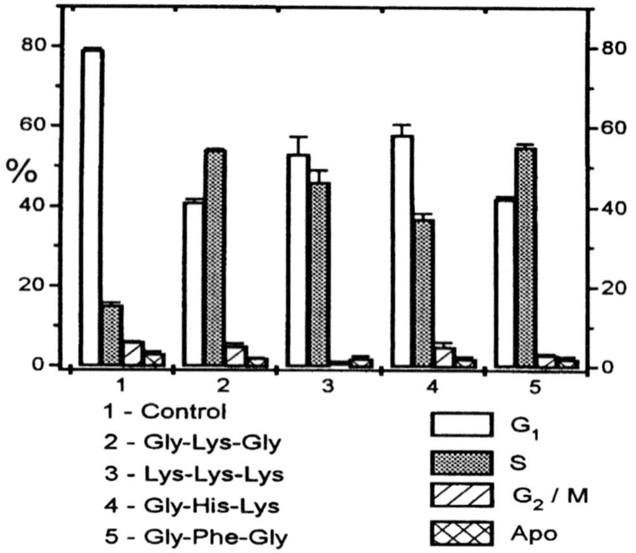

Fig. 3 Cell-cycle phases distribution on day 3 in media supplemented with various peptides. Apo – apoptotic cells

supplemented with lysine-containing peptides, as well with the control peptide Gly-Phe-Gly (Fig. 3) (Franek et al. 2003). This finding points to a complex mechanism of the action of peptides on the cultures. No correlation could be seen between the values of the monoclonal antibody yield and the proportions of the cell-cycle phases.

According to analyses performed in the experiments with the peptide Gly-Lys-Gly, the half-life time of the peptide in the culture can be estimated to about 2 days. The findings are compatible with a view that peptide molecules hit specific targets in the cells, and long-lasting alterations of cell metabolism and cell proliferation mechanism follow. These alterations seem to last for several days, and then gradually disappear.

The cell-cycle phase distribution was analyzed also in Gly-Lys-Gly – supplemented fed-batch cultures running for more than 10 days. Only at the end of this culture period did the differences between the peptide-supplemented culture and the control vanish. In both cell populations the G_0/G_1 phase cells represented 67–70%, while the S phase cells represent only 21–25%, and the G_2/M phase 7–9% of the cells (Franek et al. 2003).

Summary and Conclusions

New horizons for research and applications in animal cell culture technology were opened by our fractionation of crude peptones, and revealing the variations in the effects of peptides of different molecular sizes. We found that highly positive effects of hydrolysate fractions, the amino acid composition of which was rather unbalanced, and the essential amino acids were strongly under-represented.

These recently revealed facts led us to study the effects of individual pure synthetic peptides, because only with pure substances can the final definition and explanation of the action of peptides on cultured cells be achieved.

At present stage of the research the effects of pure peptides may be characterized by the following points:

1. The effects of peptides increase with peptide chain length up to that of pentapeptides.
2. The effects of peptide-constituting amino acids in free form are different from the effects of the peptides.
3. The population of active peptides is rather heterogeneous. The population of these peptides neither contains any common amino acid nor any common sequence motif.
4. Oligomers of non-proteinogenic D-alanine and β-alanine are active as well.
5. Some peptides exclusively increase cell density, others improve both cell density and product yield, and some peptides suppress cell growth and enhance the product yield.
6. In all cases the product yield may be enhanced by feeding the cultures with a concentrated mixture of amino acids and other nutrients.
7. The general effect of all peptides is a certain extension of high cell viability.
8. Peptides are effective at concentrations >1 mM.
9. The peptides are relatively stable in the cultures. Utilisation of liberated amino acids is marginal.
10. In the presence of peptides, the distribution of cell-cycle phases is altered.

Our results yield evidence that peptides smaller than any known peptide hormone represent a new class of agents enabling a broad spectrum of intervention with the processes in animal cell cultures. The number of sequence variants of tri- or tetrapeptides is extremely high. Only a small fraction of this peptide population could have been tested until now. Continuing the assays and selection of the most active peptides promise to lead finally to the understanding of the mechanism of action of these substances.

Pure synthetic peptides promoting cell growth and/or product synthesis would represent a basis for novel media formulations, meeting the severe criteria of biological safety. The presentation of a survival-signal to producing cells may open the way to complex feeding strategies exploiting the extended lifespan of cultures for continued product synthesis. Feeding the cultures not only with nutrients but also with the peptide is another promising possibility of improving the production of animal cell cultures.

Our studies will hopefully pave the way to the identification of peptides, enabling rational manipulation with the balance of growth versus production in animal cell technological processes.

Future Developments

The number of peptides screened so far is not sufficient to make final identification of most active compounds. It is necessary to extend the screening to test higher number of synthetic peptides. Another line of the search for active peptides may exploit the hydrolysates of proteins. High-performance fractionation procedures could be applied to obtain homogeneous peptides. These may be sequenced and the obtained knowledge combined with the knowledge gained through the screening of libraries of synthetic peptides.

The hybridoma cell lines used so far for activity assays represent a model important from the point of view of the presently existing substantial share of monoclonals in the biopharmaceutical production of therapeutics and diagnostics. Extension of the set of cell models for screening of peptides to other typical production cell lines of different tissue origin is one of the necessary tasks of the research. Preliminary studies with the CHO cell line show that the character of the effects of peptides might be analogous to the character of action on lymphocyte hybridomas.

The mechanism through which certain oligopeptides regulate proliferation of cultured animal cells is not yet elucidated. Thus the most important objectives of further research are (1) the elucidation of the mode of action of the peptides on signal transduction cascades, and (2) the identification of genes that are activated or suppressed.

Editors note Shortly after submission of this chapter, Frantisek Franek passed away. He was a pioneer in the elucidation of the mechanisms by which protein hydrolysates, peptides and amino acids increase the productivity of animal cell cultures. The field will miss his insight and devotion to this subject. We know that his approach and his contributions will be appreciated and we hope that his work will be continued by others in the field.

References

Bare G, Charlier H, De Nijs L, Verhoye F, Schneider Y-J, Agathos S, Thonart P (2001) Efects of rice protein hydrolysates on growth of CHO Cells and production of human interferon- gamma in serum-free medium. In: Lindner-Olsson E et al (eds) Animal cell technology: from target to market. Kluwer, Dordrecht, The Netherlands, pp 217–219

Fenge C, Fraune E, Freitag R, Scheper T, Schugerl K (1991) On-line monitoring of monoclonal antibody formation in high-density perfusion culture using FIA. Cytotechnology 6:55–63

Franek F (2003) Antiapoptotic activity of synthetic and natural peptides. Abstracts of the 18th ESACT Meeting, Granada, Spain, May 11–14, p. 108

Franek F (2004) Gluten of spelt wheat (Triticum aestivum Subspecies spelta) as a source of peptides promoting viability and product yield of mouse hybridoma cell cultures. J Agric Food Chem 52:4097–4100

Franek F, Dolníková J (1991) Nucleosomes occurring in protein-free hybridoma cell culture. Evidence for programmed cell death. FEBS Lett 284:285–286

Franek F, Katinger H (2001) Specific effects of synthetic oligopeptides in animal cell culture. In: Lindner-Olsson E et al (eds) Animal cell technology: from target to market. Kluwer, Dordrecht, The Netherlands, pp 164–167

Franek F, Katinger H (2002) Specific effects of synthetic oligopeptides on cultured animal cells. Biotechnol Prog 18:155–158

Franek F, Šrámková K (1996a) Cell suicide in starving hybridoma culture: survival-signal effect of some amino acids. Cytotechnology 21:81–89

Franek F, Šrámková K (1996b) Protection of B lymphocyte hybridoma against starvation induced apoptosis: survival-signal role of some amino acids. Immunol Lett 52:139–144

Franek F, Vomastek T, Dolníková J (1992) Fragmented DNA and apoptotic bodies document the programmed way of cell death in hybridoma cultures. Cytotechnology 6:117–123

Franek F, Hohenwarter O, Katinger H (2000) Plant protein hydrolysates: preparation of defined peptide fractions promoting growth and production in animal cells cultures. Biotechnol Prog 16:688–692

Franek F, Eckschlager T, Katinger H (2003) Enhancement of monoclonal antibody production by lysine-containing peptides. Biotechnol Prog 19:169–174

Heidemann R, Zhang C, Qi H, Rule JL, Rozales C, Park S, Chuppa S, Ray M, Michaels J, Konstantinov K, Naveh D (2000) The use of peptones as medium additives for the production of a recombinant therapeutic protein in high-density perfusion cultures of mammalian cells. Cytotechnology 32:157–167

Hiller GW, Clark D, Blanch HW (1993) Cell retention-chemostat studies of hybridoma cells–analysis of hybridoma growth and metabolism in continuous suspension culture on serum-free medium. Biotechnol Bioeng 42:158–195

Holdread S, Brooks W (2004) Application of peptone hydrolysates in cell culture. Bioprocess Int 2:12–13

Jan DHC, Jones SJ, Emery AN, Al-Rubeai M (1994) Peptone, a low-cost growth promoting nutrient for intensive animal cell culture. Cytotechnology 16:17–26

Mercille S, Massie B (1994) Induction of apoptosis in nutrient-deprived cultures of hybridoma and myeloma cells. Biotechnol Bioeng 44:1140–1154

Mizrahi A (1977) Primatone RL in mammalian cell culture media. Biotechnol Bioeng 19:1557–1561

Nyberg GB, Balcarcel RR, Follstad BD, Stephanopoulos G, Wang DIC (1999) Metabolism of peptide amino acids by Chinese Hamster ovary cells grown in a complex medium. Biotechnol Bioeng 62:324–335

Schlaeger E-J (1996) The protein hydrolysate, Primatone RL, is a cost-effective multiple growth-promoter of mammalian cell culture in serum-contaning and serum-free media and displays anti-apoptotic properties. J Immunol Meth 194:191–199

Singh RP, Al-Rubeai M, Gregory CD, Emery AN (1994) Cell death in bioreactors: a role for apoptosis. Biotechnol Bioeng 44:720–726

Vomastek T, Franek F (1993) Kinetics of development of spontaneous apoptosis in B cell hybridoma cultures. Immunol Lett 35:19–24

Chapter 6
Use of Protein Hydrolysates in Industrial Starter Culture Fermentations

Madhavi (Soni) Ummadi and Mirjana Curic-Bawden

Abstract Lactic acid bacteria (LAB) have been used as starter cultures for fermenting foods long before the importance of microorganisms were recognized. The most important group of LAB are the lactococci, lactobacilli, streptococci, and pediococci. Additionally, bifidobacteria have been included as a probiotic, providing added value to the product. Since the genera involved are so diverse, the nutritional requirements (energy, carbon and nitrogen sources) differ significantly between and within species. Designing an optimum fermentation medium for production of active and vigorous LAB starter cultures and probiotics requires selecting the right raw ingredients, especially protein hydrolysates that can provide adequate nutrients for growth and viability. This chapter attempts to describe the application of various commercial protein hydrolysates used for production of dairy and meat starter cultures, with special emphasis on meeting the nitrogen requirements of industrially important LAB species.

Keywords Starter cultures • Protein hydrolysates • Fermentation growth medium • Nutritional requirements of lactic acid bacteria

Abbreviations

Ala	Alanine
Arg	Arginine
Asn	Asparagine
Asp	Aspartic acid (Aspartate)
CDM	Chemically defined medium

M. (Soni) Ummadi (✉)
Dreyer's Grand Ice Cream, Bakersfield, CA, USA
e-mail: Soni.Ummadi@dreyers.com

M. Curic-Bawden
Chris Hansen, Inc., Milwaukee, WI, USA

V.K. Pasupuleti and A.L. Demain (eds.), *Protein Hydrolysates in Biotechnology*,
DOI 10.1007/978-1-4020-6674-0_6,

Cys	Cysteine
Gln	Glutamine
Glu	Glutamic acid (Glutamate)
Gly	Glycine
His	Histidine
Ile	Isoleucine
Leu	Leucine
Lys	Lysine
Met	Methionine
Phe	Phenylalanine
Pro	Proline
Ser	Serine
Thr	Threonine
Trp	Tryptophan
Tyr	Tyrosine
Val	Valine

Introduction

Microorganisms are employed in the manufacture of a wide range of fermented products. Fermented foods were manufactured long before the importance of microorganisms was recognized. However, the present knowledge of growth of microorganisms and their metabolism permits us to manufacture these fermented products that are uniform, consistent in composition, and predictable in flavor.

Current trends in the food industry focus on implementing standardized, economical, and reproducible production processes, which ultimately result in standard quality products. This trend influences the quality requirements of various food ingredients, including starter cultures.

Starter cultures are currently used in a wide range of food applications such as in the preparation of fermented milks (yogurt, buttermilk), fermented cream, a variety of soft and hard cheeses (Cheddar, Swiss, Gouda, Mozzarella, etc.), fermented meats and sausages, sourdough and fermented vegetables (pickled vegetables, sauerkraut etc.) (Table 1).

Commercial lactic starter cultures are produced in such a way that when used in their final product application, they produce the desired effect. It is of utmost importance to conduct well-controlled fermentation research for determining the best-possible formulation. Every strain has its distinctive growth requirements for essential energy, carbon and nitrogen sources. The choice of the right raw ingredients and the formulation of the fermentation medium is undoubtedly the most important step in the production of these concentrated starter cultures.

The nutritional requirements of lactic acid bacteria (LAB) starter cultures differ significantly and are correlated to differences between and within species. The growth of these nutritionally fastidious strains (especially Lactobacilli) is highly dependent on the type of carbon and nitrogen sources available in the fermentation medium.

Table 1 Bacterial species commonly used as starter and flavor adjunct cultures and their applications

Organism	Culture area	Food applications
Lactococcus lactis *Leuconostoc* spp.	Mesophilic dairy cultures	Fermented milk Sour cream Cream cheese Soft cheeses Cheddar cheese Continental type cheeses
Streptococcus thermophilus	Thermophilic dairy cultures	Yogurt, semi-hard cheeses
Lactobacillus spp.	Thermophilic dairy cultures	
• *L. bulgaricus*		Yogurt, cheese
• *L. helveticus*		Fermented milk, cheeses
• *L. casei*		Cheeses
• *L. acidophilus*		Fermented milk, probiotics
• *L. reuteri*		
Bifidobacterium spp.	Thermophilic dairy cultures	Fermented milk Probiotics

Industrial fermentations are usually performed in semi-defined and complex media, which use less expensive raw materials that can yield higher levels of biomass or fermentation derived products. Crude and unidentified additives such as peptones, protein hydrolysates, yeast extracts, growth factors etc., can be added to the fermentation media, to provide a broad spectrum of nutrients. Protein hydrolysates from various origins are known to have different effects on various strains of LAB; while some may support growth of an organism, others may induce a specific type of amino acid metabolism (Loader et al. 1999). Therefore, it is important to study the effect of various protein hydrolysates on lactic starter cultures.

Extensive research has been conducted in last 3 decades to elucidate the nutritional requirements and metabolism of specific strains but the knowledge cannot be easily correlated into industrial fermentations. Growth medium in most cases has to be tailored for each individual strain. Also, in many instances, the information is retained as a trade secret.

This chapter attempts to describe the application of various protein hydrolysates that can be used during fermentations, and also focuses on the formulation of the growth medium for commonly used LAB strains with a special regard to nitrogen requirements. Typical compositions of various types of commercial protein hydrolysates will also be presented.

Industrial Applications of Starter Cultures

A wide variety of organisms are used as starter cultures (Table 1). When applied in food fermentations, starter cultures have different roles based on their specific metabolic characteristics. The primary role of starter cultures is to convert the available source of carbohydrates (e.g., lactose in milk) into acid (mainly lactic acid), lower pH, and thus preserve the product and contribute to its characteristic taste and flavor. Some strains of *Lactobacillus acidophilus*, *Lactobacillus casei*, *Lactobacillus reuteri* and *Bifidobacterium* species are known to have a probiotic effect. These organisms are used to complement primary starter cultures and contribute to the increased value of the final fermented product.

In order to meet the required metabolic activities (e.g., acidification rate), starter cultures have to be inoculated in sufficient amounts and have cells that are vigorous and healthy. This requires a well-designed and carefully controlled process of culture production.

Production of Starter Cultures

Commercial starter cultures are available as single or multiple strains, and as defined or undefined cultures. Development of the range and quality of starter cultures is driven by the demands posed by the food industry. The current trend emphasizes the application of concentrated frozen or freeze dried starter cultures.

Commercial starter culture concentrates are either available as "bulk sets" (Redi set) or "direct vat sets" (DVS). "Bulk sets" are used to prepare intermediate starter cultures, which are then inoculated into the production vat to prepare a final product. These cultures are available in frozen (70 mL) or freeze-dried form (5–10 g package) and are designed to inoculate 100–1,000 L of material. Unlike bulk cultures, DVS cultures act as a direct inoculum in the final product. They are available as frozen or freeze-dried cultures where ~500 g of frozen culture is used to inoculate 2,500–5,000 L of material, depending upon the culture type and application.

As mentioned above, one of the primary requirements in the production of a microbial starter culture is to obtain high yield of vigorous biomass. The "yield" and "quality" of biomass is of utmost importance, and this depends upon the production procedures (configuration of the equipment, design of the fermentation process, medium composition and growth conditions), downstream processing (concentration, addition of cryoprotective agents; freezing or freeze-drying parameters) and storage conditions. More detailed information on production of different types of starter cultures is captured in articles by Porubcan and Sellars (1979) and Hoier et al. (1999).

Fermentation specialists are constantly looking into a variety of ways to produce higher biomass concentrations. However, during production of starter cultures, the focus should not only be on biomass quantity (cell yield) but rather on biomass quality. One of the most important parameters of the biomass quality is the composition and status of the cell wall of the organism, which determines the survival rate of cells during downstream processing. Even though the robustness of the cell wall is species and even strain-dependent, it will also be influenced by the composition of the medium, growth conditions, and physiological status of cells at the time of harvesting. Presence or absence of certain nutrients and co-factors, can have an effect on the cell wall thickness, elongation of cells, and cell division directly affecting the strain survival rate. Therefore, the production procedure has to be optimized for each culture/strain depending on the characteristic traits of the organism, growth requirements, and final application of the culture.

Design of Fermentation Media

Detailed understanding of microbial metabolism is a pre-requisite in optimizing a fermentation medium to achieve high quality biomass, which in turn will maintain the metabolic activity after production and downstream processing (i.e., concentration and freeze drying etc.).

Primarily, fermentation media have to fulfill the complex nutritional requirements of each specific species/strain, e.g., provide essential elements such as amino acids, peptides, vitamins, minerals, nucleic acid bases and other growth factors. Table 2 highlights some commonly used raw ingredients grouped after their primary function. To be used in fermentations, raw ingredients need to be inexpensive, readily available, and of reproducible quality. Unfortunately, many of the complex nutrients listed in Table 2 are subject to large inconsistencies due to the nature of biological material, and production of these materials in different regions and seasons. These variations can significantly impact the quality of the final product, resulting in significant

Table 2 Ingredients used for the preparation of media for commercial starter production (Modified from Champagne 1998)

Type or function	Ingredient
Sugars	– Monosaccharides (glucose, fructose)
	– Disaccharides (lactose, sucrose, maltose)
	– Dextrins and maltodextrins
	– Non fat dry milk (lactose)
	– Whey (lactose)
Proteins	– Non fat dry milk
	– Whey
	– Whey protein concentrates
Other nitrogen sources	– Peptones
	– Casein hydrolysates
	– Whey protein hydrolysates
	– Soy protein hydrolysates
	– Meat protein hydrolysates
	– Hydrolyzed cereal solids
	– Yeast extracts
Vitamins and minerals	– Yeast extracts
	– Corn steep liquor
Others	– Tween/oleic acid
	– Mineral salts
	– Defoamers
	– Buffers

production losses. An often-practiced strategy to overcome this problem is to use blends of the same ingredient-type but from different manufacturers.

Energy and Carbon Sources

LAB do not possess a functional tricarboxylic acid (TCA) cycle, which makes their energy generating pathways inefficient. Homofermentative organisms such as *L. lactis, S. thermophilus*, *L. bulgaricus*, *L. helveticus*, and *L. acidophilus* generate energy via the glycolytic (Embden-Meyerhoff- Parnas-EMP) pathway, where two moles of ATP are formed per mole of hexose consumed. *Leuconostoc* sp. generates energy via a heterofermentative pathway and only one mole of ATP is formed per mole of hexose consumed. Additionally, ATP may be generated by a chemi-osmotic energy process e.g. lactate efflux (Konings et al. 1989). LAB in general does not produce any endogenous energy storage compounds such as glycogen, polyphosphate, and poly-β-hydroxybutyrate, except for the insignificant amounts of the phosphoenol pyruvate pool. One exception is *Bifidobacterium bifidum*, which during the stationary phase forms storage compounds such as glycogen and polyphosphates (Norris et al. 1954).

Therefore, in order to support anabolic processes and cell growth, energy sources have to be supplied in sufficient amounts. The preferred energy sources for most industrial LAB fermentations are lactose and glucose. During growth in a complex fermentation medium that contains peptides or amino acids, the main function of the sugar source is to generate energy. However, if the cell wall ingredients (teichoic acid and peptidoglycans) and RNA precursors (purine and pyrimidine bases) are not supplemented into the medium, a small fraction of sugar and a portion of ATP will be used for these purposes. It is therefore important to supplement adequate forms of nitrogen sources and growth factors.

Nitrogen Requirements and Uptake Systems in LAB

The availability of nitrogen sources (proteins, peptides and amino acids) is essential for the growth of starter cultures. Sources of amino acids are peptides found in milk and in other hydrolyzed proteins (Foucaud et al. 2001). LAB have a limited capacity to synthesize amino acids, hence they depend on exogenous sources of amino acids and peptides.

In order to use available proteins, peptides, and/or free amino acids as building blocks for synthesis of new proteins and enzymes, they have to be translocated into the cell. If large proteins and peptides cannot be handled by the uptake system of the cell, these proteins and peptides have to be further hydrolyzed. Some LAB that can synthesize and secrete extracellular proteases, break proteins down into peptides and amino acids, making them available for translocation. In a natural environment to utilize milk proteins, a *L. lactis* strain depends on a proteolytic system consisting of a cell-envelope-located proteinase, three peptide transport systems, a set of intracellular peptidases, and nine different amino acid transport systems, which act in concert to supply the cell with essential and growth stimulating peptides and amino acids (Juillard et al. 1996, 1998; Kunji et al. 1996; Mierau et al. 1997). Conversely, LAB that lack proteolytic systems have to rely on basic building blocks such as amino acids and ammonia as sources of nitrogen.

An alternative and a more economical option with regards to consumption of cell energy, is an application of already digested protein hydrolysates. Digestion of a protein source results in the release of amino acids and peptides to a different degree of hydrolysis (DH%). The size and amino acid composition of these released peptides influence the growth rate and biomass yield of the organism. However, not every amino acid that is released would impart the same biological value and growth promoting effect to the bacterial cell. Only amino acids in an L-form can be incorporated into the cell biomass.

As mentioned above, in order to be used as a source of amino acids, the available peptides have to be successfully translocated into the cell. Oligopeptides represent a main source of nitrogen for *L. lactis* during its initial growth phase in milk (Juillard et al. 1995). Juillard et al. (1998) observed that translocation of oligopeptides is controlled by two main factors: the isoelectric point of the peptide and its molecular weight (MW). Substrate specificity studies conducted by Juillard et al. (1995, 1998) on *L. lactis* MG1363 oligopeptide transport systems indicated that the size and the form of oligopeptides clearly determine their translocation into the cell.

When compared to free amino acids, small peptides are of superior nutritional value, because of their ability to transport more than one amino acid into the cell at the expense of one ATP. This is considered to be much more "economical" for the bacterial cell. The advantage of direct transport of peptides (two to three units) into the cell lies in the reduction of the amount of metabolic energy used for amino acid uptake. On the other hand, initial growth of *L. lactis* strains was supported by larger oligopeptides (Juillard et al. 1995). Hence, the optimal solution would be a balanced combination of oligopeptides and di- and tri-peptides.

Protein Hydrolysates: A Source of Nitrogen and Growth Factors

There are nearly 100 different types of commercially available hydrolyzed proteins produced from inexpensive sources ranging from vegetables to meat and milk. Examples include caseins, whey proteins, skim milk, milk and whey protein isolates and concentrates, lactalbumin, meat and animal tissue, collagen, gelatin, corn, cottonseed, soy meal and soy protein isolates and concentrates. Some of these materials are by-products of other processes and are intrinsically variable. Even though a particular source of protein (e.g. casein and whey) would have a fairly constant composition of amino acids, various protein hydrolysates from the same source still have significant variations in peptide and amino acid profiles (Table 3).

Table 3 Examples of various commercial hydrolysates based on dairy proteins (Adapted from Commercial Product Information Bulletins)

Supplier	*Quest*	*Quest*	*Sensient*	*Sensient*	*DMV*	*Sensient*	*DMV*
	Hy-Case Amino – Acid digest of casein	*NZ-Amine EKC – Enzyme digest of casein*	*EHC-K – Enzyme digest of casein*	*AMF 2234 – Enzyme digest of casein*	*CE90MJ-K Enzyme digest of casein*	*Pancase S – Pancreatic digest of casein*	*CE90MS Pancreatic digest of casein*
AN/TN ratio	69.4	24.2	26.0	20.0	33.0	34.0	49.2
A Acids	*Total (mg/g)*	*Total (mg/g)*	*Total (mg/g)*	*Total (mg/g)*	*Total (mg/g)*	*Total (mg/g)*	*Total (mg/g)*
ALA	22.3	19.1	25.2	23.6	26.0	23.7	22.0
ARG	24.1	33.3	29.1	30.7	29.0	32.2	27.0
ASP	45.7	64.8	55.2	59.7	57.0	60.6	71.0
CYS	0.2	2.9	0.6	0.4	4.0	0.2	3.0
GLU	135.0	190.2	171.8	181.2	181.0	189.3	177.0
GLY	11.8	16.4	14.8	14.8	15.0	15.0	18.0
HIS	11.0	24.6	22.2	24.8	20.0	23.0	16.0
ILE	26.9	48.8	50.8	52.5	45.0	54.9	39.0
LEU	45.3	80.1	78.2	78.7	75.0	82.6	66.0
LYS	52.7	73.1	62.0	69.2	63.0	67.3	65.0
MET	19.9	24.1	2.0	16.6	23.0	4.7	18.0
PHE	36.0	47.3	39.6	39.6	41.0	41.5	31.0
PRO	58.4	88.2	52.8	44.1	87.0	39.3	74.0
SER	34.5	51.0	34.7	38.6	39.0	45.2	38.0
THR	21.6	34.0	29.4	31.3	31.0	36.5	27.0
TRP	<0.5	9.5	0.0	0.0	8.0	0.0	5.0
TYR	12.3	30.6	29.1	20.0	49.0	9.9	8.0
VAL	37.5	58.5	55.2	53.8	54.0	56.2	48.0

On the other hand, protein sources such as casein, meat, gelatin, and soy, etc. differ from each other in their total amino acid composition and protein structure. While comparing dairy peptones to meat/soy hydrolysates, one notices that dairy hydrolysates are abundant in amino acids such as Glu, Ile, Leu, Val, and Met, whereas meat/soy hydrolysates are superior sources of amino acids namely Arg, Cys, Gly, and Pro (Table 4). However, it is not simply the total levels of amino acids but also their form (i.e. oligopeptides, di- and tri-peptides, or as free amino acids) that determines the biological value of the particular protein hydrolysate.

Regardless of the protein source, the various hydrolyzing agents that can be currently used, gives the manufacturer an opportunity to design a protein hydrolysate with a range of peptide fractions and desired ratio of peptides to free amino acids. The release of peptides, size of peptides highlighted as peptide profile (Fig. 1), the ratio between amino nitrogen when compared to total nitrogen (AN/TN ratio) and the levels of free amino acids (Figs. 1 and 2) are determined by the type of digestion process (acid versus enzyme hydrolysis); choice of enzymes (purified or blends of animal/microbial enzymes) that have different enzyme specificities; and also by altering processing parameters (pH, temperature and incubation/processing time). Based on the degree of hydrolysis (DH%), the commercial protein hydrolysates can have different total protein levels and different AN/TN levels.

The effect of various hydrolyzing agents on casein as illustrated in Fig. 1 highlights the degree of hydrolysis and the peptide and amino acid distribution profiles. Hydrolysates with low molecular weight peptides are known to stimulate growth of dairy lactic acid bacteria. Also highlighted are differences in levels of free amino acids released by the action of two pancreatic digests attributing the changes to the hydrolyzing conditions used with each treatment.

When considering using a commercial protein hydrolysate, it is important to note the AN/TN ratio. Table 3 highlights the AN/TN ratios and amino acid profiles of various commercial casein hydrolysates. According to the product bulletin literature, the AN/TN ratio with the enzyme-hydrolyzed caseins ranges between 13% and 30% and with pancreatic digest to be between 28% and 49% with acid hydrolyzed caseins to be 60–80% (Table 3). Acid hydrolysis of casein is not site specific and therefore results in more uniform size of peptides and higher levels of low MW peptides (Fig. 1). Depending on the processing time and conditions applied, it can completely release all free amino acids (e.g., Casamino acids). As a result of the production process, acid hydrolyzed proteins may contain up to 40% of NaCl. Salt free preparations (e.g. Hy-case SF by Quest) are available at an increased cost. In enzymatic and acid hydrolysis, certain amino acids e.g. cysteine (Cys) in casein peptones and tryptophan (Trp) in acid-digested casein can be denatured to different levels making it unavailable for bacterial utilization (Fig. 2). The data shown in Table 3 for total amino acids (mg/g) present in the hydrolysates is useful but to a fermentation specialist, to know if they are available as free or as peptide bound amino acids would be even more useful. Additionally, when looking into the product information, it is important to ask the vendor for

Table 4 Comparison of dairy, meat and soy protein hydrolysates (Adapted from Commercial Product Information Bulletins)

Supplier	*DMV*	*Quest*	*DMV*	*Sensient*
	SE50 MK Kosher Enzyme digest of soy	*Primatone G Enzyme digest of gelatin*	*AED80M Enzyme digest of animal tissue*	*AMF 2234 Enzyme digest of casein*
AN/TN ratio	26.7	17.3	23.4	20.0
Am Ac	*Total (mg/g)*	*Total (mg/g)*	*Total (mg/g)*	*Total (mg/g)*
ALA	25.0	84.5	68.0	23.6
ARG	40.0	77.7	48.0	30.7
ASP	61.0	62.1	61.0	59.7
CYS	8.0	1.6	5.0	0.4
GLU	104.0	97.9	100.0	181.2
GLY	24.0	194.0	125.0	14.8
HIS	13.0	8.2	24.0	24.8
ILE	25.0	14.0	NA	52.5
LEU	44.0	29.1	56.0	78.7
LYS	36.0	38.0	52.0	69.2
MET	7.0	10.1	14.0	16.6
PHE	29.0	18.8	33.0	39.6
PRO	30.0	122.0	78.0	44.1
SER	27.0	37.4	25.0	38.6
THR	23.0	18.2	21.0	31.3
TRP	3.0	<0.5	3.0	0.0
TYR	19.0	3.7	19.0	20.0
VAL	28.0	24.8	36.0	53.8

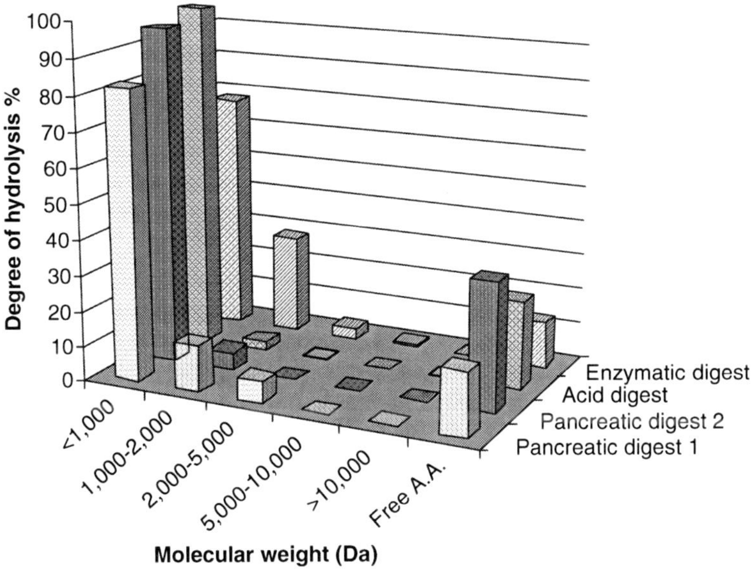

Fig. 1 Peptide distribution in casein hydrolysates digested using different enzymes and/or hydrolyzing treatments (Adapted from Commercial Product Information Bulletins)

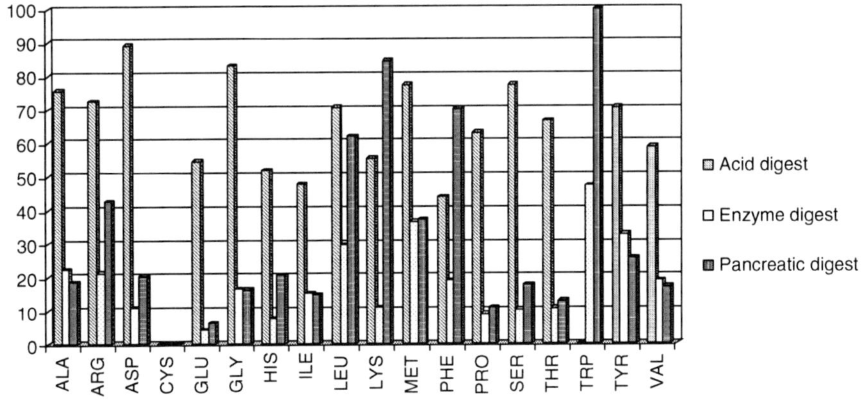

Fig. 2 Effect of digestion treatment on levels of free amino acids in casein hydrolysates (Adapted from Commercial Product Information Bulletins). X-axis represents the amino acids and the y-axis represents the % free amino acids

the methods used for analysis and quantitation of these amino acids and also if they are comparable methods.

Commercial protein hydrolysates are available in different levels of purity, ranging from technical grade to refined, food or non-food grade, and can also be available as Kosher. In production of starter cultures only food grade materials are used. Table 4 compares the AN/TN ratios and total amino acid profiles of dairy, meat and soy protein hydrolysates. This gives useful information in terms of the total amino acids present in that particular hydrolysate. However, if Kosher or Kosher for Passover status is a customer requirement, then the use of meat and a few soy proteins is prohibited and the fermentation specialist should look into a protein hydrolysate that would provide the most amino acids and peptides that are derived from dairy or other vegetable origins.

To complement protein hydrolysates and supplement nutritive compounds and growth factors, yeast extracts are also widely used in fermentation growth media. Yeast extracts are concentrates of soluble components of yeast cells. They are produced either by autolysis – hydrolysis by their own endogenous enzymes or by more controlled digestion by added proteases (mixtures of proteases and peptidases). Salt free yeast extracts contain 73–75% of protein, of which 35–40% are free amino acids; 10–15% are di- and tri-peptides of MW <600 Da and 40–45% are oligopeptides of MW 2,000–3,000 Da (Summer 1996). Besides being an excellent nitrogen source, yeast extracts are also a source of nucleotides (RNA and DNA precursors), vitamins and trace elements such as Ca, Mg, Mn, and Fe. More aspects about the selection of peptides with specific molecular weights will be presented later on in this chapter.

Criteria for Selection of Protein Hydrolysates

Protein hydrolysates need to be selected based on the nutritional needs of the microorganism of interest, i.e., their need for peptides, amino acids and other nutrients. In addition, other fermentation parameters such as providing adequate carbon, vitamin, minerals and growth factors enable predictable fermentation times, cell yield and viability through downstream processing and shelf life storage.

The criteria of selecting a protein hydrolysate depend on the following factors:

1. *Nutritional requirement of the strain* – Degree of hydrolysis, peptide profile, ratio between amino nitrogen (AN) and total nitrogen (TN), level of free amino acids, and mineral content.
2. *Cost of the protein hydrolysates* – quality of raw ingredients used, ease of biomass recovery from the medium, and price-performance ratio.
3. *Regulatory compliance* – In order to be used in the fermentation media for production of starter cultures, hydrolyzed proteins have to be food grade.
4. *Additional market and/or customer requirements* – In some cases, additional requirements in respect to the raw material need to be fulfilled, such as Kosher, dairy only (i.e. milk protein based); GMO free (i.e. soy free); and dairy free (i.e. soy based hydrolysates), etc.

Nutritional Requirements and Use of Protein Hydrolysates During Fermentations

Since this chapter focuses on lactic acid bacterial starter cultures, the discussion below is dedicated mainly to nutritional requirements and fermentation media for production of these species and strains.

Genus Lactococcus

Lactococci are used as main components of dairy cultures applied in the production of a wide variety of hard and semi-hard cheeses, fermented milks and cream. Strains that are now used as starter cultures were once isolated from dairy and plant origins and selected based on their specific metabolic properties. Due to their long adaptation to milk, which is a fairly nutritious growth medium, dairy strains, over the years have evolved and lost their strong proteolytic activity characteristics, and acquired auxotrophy to most amino acids (Bringel and Hubert 2003; Morishita et al. 1981).

Amino acid requirements and transport systems are known to be growth-limiting factors in lactococci (Poolman and Konings 1988). Dairy *Lactococcus lactis* subsp. *lactis* is auxotrophic for at least 7 amino acids: Gln, Met, Leu, Ile, Val, Arg and His (Law et al. 1976). According to Hugenholtz et al. (1987) the specific growth rates of several *Lactococcus lactis* subsp. *cremoris* strains (HP, E8, ML1 and AM1, NIZO collection) were 10–40% lower in milk than in growth media supplemented with amino acids or caseins. Increased hydrolyzed casein levels up to 4% (w/v) or addition of amino acids (specifically Leu and Phe and Glu) did indeed increase the specific growth rate of lactococci by 10–20%.

Even when a lactococcal strain possesses a strong proteolytic system, growth on specific proteins, such as β-casein is reported to be limited due to low levels of free amino acids, namely His, Leu, Gln, Val, and Met (Kunji et al. 1995). Requirements for Glu and Phe are strain dependent (Henrik Moellgaard, CH-internal communication). *L. lactis ssp. cremoris* strains are generally more demanding than L. lactis ssp. lactis and often require additional Tyr, Asn and Ala. Also, when essential amino acids are being released too slowly from their corresponding peptides to support optimal growth, need for protein hydrolysates become necessary (Hellinick et al. 1997).

In the last 2 decades, numerous chemically defined media were developed with the aim of investigating metabolic and energy requirements of *L. lactis*. The standard synthetic medium (MCD), which contains 47 components, including 18 amino acids and 14 vitamins, was initially developed by Otto et al. (1983) and modified by Poolman and Konings (1988). Omission of non-essential amino acids from the chemically-defined medium resulted in 75% lower growth rate and up to 50% lower biomass yield (Novak et al. 1997).

Proline (Pro) was found to stimulate growth of *L. lactis* strains, regardless of the ability to synthesize this amino acid (Smid and Konings 1990). Milk is abundant in Pro, but this amino acid is not readily available, since some strains lack a proline transport system. Proline can only be transported into the cell by passive diffusion or by transport of proline containing peptides (Smid and Konings 1990). Hence, it is not just the level of proline as free amino acid (Fig. 2), but also its form, which should be taken into consideration when selecting a suitable protein hydrolysate for the growth medium of *L. lactis*.

Law et al. (1976) compared the effect of whey, papain digest of casein, and two commercially available soy peptones on growth of four lactococcal strains and found that the soy peptones were a much more preferable nitrogen supplement than whey or papain digest of casein. Further analysis indicated that soy peptones, when supplemented at 2.5% (w/v) into the medium, contained a greater proportion of small peptides (three to seven amino acid residues) and free essential amino acids than the papain digest of casein, which probably were responsible for stimulating the growth of these four *L. lactis* strains (Law et al. 1976).

To study the substrate specificity of *L. lactis* MG1363, Juillard et al. (1998) studied chemically-defined medium supplemented with low MW milk peptides (<1,000 Da) or a tryptic digest of αs_2-casein as a source of amino acids. The results indicated a preferential use of milk peptides which were hydrophobic and contained basic peptide fractions with molecular masses ranging between 600–1,100 Da, and an inability to use large, acidic phosphopeptides with MW between 1,464–3,152 Da. This was further explained by substrate specificity of the oligopeptide transport system.

An option that seems to work for many *L. lactis* strains is the use of enzyme hydrolyzed caseins with usage levels between 0.5% and 2.5%. Keeping in mind the cost of these hydrolysates, a concerned fermentation specialist may use several commercially available proteolytic enzymes with various specificities to pre-digest milk proteins prior to using them as nitrogen source.

Genus Lactobacillus

Several lactobacilli species are used as components of starter cultures for the production of yoghurt (*L. delbrueckii* ssp. *bulgaricus*), and various types of cheeses (*L. helveticus, L. paracasei*). These are of great industrial importance in the dairy/cheese industry. Beside these, selected strains of *L. acidophilus, L. johnsonii, and L. reuteri* are brought into the limelight, due to their probiotic properties. Probiotic strains are used either as components of starter cultures for fermented milks or as nutraceutical products for which they are prepared in the form of free or encapsulated freeze-dried materials.

Nutritional and nitrogen requirements vary significantly from one species to another and even between strains of the same species/sub-species. Several studies have been performed with the aim to elucidate general nutritional requirements of

Lactobacillus sp. Elli et al. (2000) and Chervaux et al. (2000) described the nutrient requirements of 22 *Lactobacillus* strains using a chemically defined medium, which contains 21 amino acids, and other nutrients comprising of 60 components. In general, for optimal growth and viability, these lactobacilli required fermentation media supplemented with abundant carbon and nitrogen sources, vitamins, micro and macronutrients and nucleotide bases.

L. helveticus strains have complex requirements for growth. The requirements for amino acids are greater than for other lactobacilli or lactococcal strains. Morishita et al. (1981), indicated that the strain ATCC15009 is auxotrophic for 14 amino acids, four vitamins and uracil while strain CRL1062 requires 13 amino acids (Hebert et al. 2000). These multiple auxotrophies of lactobacilli have been related to a single mutation in RNA polymerase (Morishita et al. 1974) or mutations in the amino acid biosynthetic pathways (Morishita et al. 1981). Therefore, the growth rates of both strains were affected by the nitrogen source present in the culture medium. When a chemically defined medium (CDM) was supplemented with low molecular weight peptides (LMWP < 3,000 Da) in a casitone-pancreatic digest of casein, the cell numbers increased by 1.3 fold. Similarly, when the same strains were grown in CDM minus amino acids supplemented with casitone as the sole amino acid source, specific growth rates were similar to those found in CDM with casitone. Also, interestingly, no growth was observed when the casitone was substituted with β-casein or casamino acids (Hebert et al. 2000). Furthermore, unlike lactococcal strains, di and tri-peptides highlighted by Juillard et al. (1998) did not influence growth of *L. helveticus* strains. The above data highlights the fact that the organism does need specific amino acids and peptides for its optimal growth.

Lactobacillus acidophilus (LA) claimed for its probiotic properties requires the presence of Pro, Arg and Glu for growth (Morishita et al. 1981) but is greatly stimulated by almost all 18 amino acids. Since, LA does not possess a fully functional pentose phosphate pathway, it requires the presence of aromatic amino acids and His too (Hebert et al. 2000). However, Thr, Ala, Asp and Asn are not considered essential for growth of LA (Henrik Moellgaard, CH-internal communication), which indicates that the amino acid precursor oxaloacetate is available in sufficient quantities for *de novo* synthesis from sugars or citrate or other supplied amino acids. Even though it is not considered essential, Arg seems to stimulate growth of tested CH-strains *L. bulgaricus L. acidophilus, L. reuteri, Pediococcus pentosaceus, S. thermophilus* (Henrik Moellgaard, CH-internal communication).

Numerous literature citations described above draw attention to the nutritional requirements of lactobacilli, but rarely do studies describing the positive effects of specific protein hydrolysates on growth and survival of LAB. Even though the following studies were not performed in fermentation media, but in milk, they contain information on effects of casein or whey hydrolysates during fermentation in fermented food applications, indicating that hydrolysates might play a beneficial role in producing starter cultures.

The effect of 2% acid casein hydrolysate and cysteine on the viability of LA strain MJLA1 in milk substrate was studied by Ravula and Shah (1998). The viability of the strain was at 10E+05 CFU/g after 12 weeks in a frozen dairy dessert compared to <10E2 cfu/g with the control sample suggesting that possibly, the nutrients such as peptides and amino acids supplied through the casein hydrolysate and alternatively cysteine help in maintaining the redox-potential and improved strain viability.

The effect of cysteine, acid hydrolyzed casein (ACH), whey powder (WP), whey protein concentrate (WPC) or tryptone (tryptic digest of casein) on the viability of *S. thermophilus*, LA, and *Bifidobacteria* sp. supplied as ABT culture (Chr. Hansen Pty. Ltd., Bayswater, Australia) in fermented dairy products were investigated by Dave and Shah (1998). The results showed yogurt mixes supplemented with ACH, WPC, and tryptone significantly increased growth and subsequently decreased the fermentation times to reach pH 4.50 while WP did not.

Gomes and Malcata (1998) determined the effect of two milk protein hydrolysates prepared from commercial proteases (*Aspergillus* sp. Protease 2A and *B. subtilis* Neutrase) on growth of LA strain Ki (NIZO, Ede,The Netherlands). The growth rate and acid production of strain Ki were not positively affected by the addition of either milk protein hydrolysate. Although, strain Ki grew slowly, its proteolytic system was apparently able to generate its own nitrogen source. On the other hand, co-culture with *Bifidobacterium lactis* (1:1 ratio) led to enhanced rates of growth and acidification when compared to the single strain suggesting that growth factors other than peptides were growth-limiting in the medium.

Lactobacillus reuteri, among the lactobacilli, are known to be the most fastidious organisms. Amino acids Met, Glu, Tyr, Trp, His, Leu, Val, and Ala are absolute requirements and the lack of other amino acids leads to decreased growth rate (Henrik Moellgaard, CH-internal communication).

Several *Lactobacilli* species (*L. johnsonii, L. gallinarum, L. gasseri, L. helveticus*) are not able to synthesize purines and pyrimidines *de novo* and the need for additional iron was dependent on the presence of DNA and RNA precursors (Elli et al. 2000). On the contrary, *L. casei, L. paracasei,* and *L. plantarum* are able to grow in the absence of purines and pyrimidines. This fact is important since milk and milk-derived hydrolysates do not contain purine and pyrimidine precursors. Therefore, when selecting the raw materials for fermentations of lactobaclli, these precursors have to be supplemented via an alternative source such as yeast extract.

With certain lactobacilli, soy protein was better utilized when it was delivered as a hydrolysate than a whole soy protein (Hsieh et al. 1999). When commercial soy protein hydrolysates (NZ Soy-BL, Hy-soy, and Amisoy) were used as a nitrogen source, the MW of soy peptides (~700 Da) had a significant influence on the production of lactic acid. Supplementation rate of ~3% (w/v) of 700 Da soy peptides led to increased growth and concentration of lactic acid to 51 g/L, while peptides in unhydrolysed protein (MW > 10,000 Da) resulted in poor growth (Hsieh et al. 1999). These results demonstrate that similar to casein hydrolysates, the MW of soy protein hydrolysates also plays a crucial role in growth and formation of lactate by lactobacilli and may be related to peptide transport mechanisms.

There are several enzyme specificities that should be taken into account when designing a fermentation process (including other media ingredients) for the lactobacilli strains that will be used as starter or adjunct cultures versus strains that will be used as probiotics. For example, the biosynthesis of lactobacilli PrtB/PrtH proteinase was shown to be regulated by the peptide pool of the culture medium and repressed by a peptide-rich medium (Kenny et al. 2003). With *L. helveticus,* enzymes such as aminopeptidases, dipeptidase, tripeptidase and endopeptidase activities are known to be strain and species dependent. Hebert et al. (2000) have shown that pepN aminopeptidase activity of *L. helveticus* CRL 1062 and pepX, pepI and pepN activities in *L. bulgaricus* (Morel et al. 1999) were independent of the peptide content of the growth medium, while pepX and pepN of lactoccocci were found to be regulated by the peptide content of growth medium. Hence, in case of the lactobacilli strains that will be applied as a cheese starter (or) adjunct culture, it is very important to select medium ingredients that are able to induce and maintain important metabolic pathways such as extracellular and intracellular proteolytic systems. These microbial traits will enable acidification of milk at standard established time and/or release of enzymes needed for the development of flavor during cheese ripening.

In contrast, the main focus in the production of probiotic strains (*L. acidophilus, L. johnsonii, L. reuteri*) is on high yield and quality biomass that will enable the cultures to survive freeze-drying and exposure to prolonged shelf life at room temperatures. In any case, choosing an appropriate protein hydrolysate to meet all the above contingent requirements for each strain becomes essential.

Genus Streptococcus

S. thermophilus (ST) strains are essential components of yoghurt cultures and some cheese cultures. Besides lactic acid, some strains produce exopolysaccharides (EPS), which contributes to the texture of fermented milk and can improve yield in some types of cheeses (Petersen et al. 2000). Hence, research efforts have been mainly focused on application of ST strains, genetics and metabolic activities related to formation of EPS. Their nutritional requirements were not investigated to the same extent when compared to lactococci.

For growth of *S. thermophilus*, the cells require a nitrogen source from the medium. Milk itself contains a potentially abundant supply of nitrogen for cell growth. However, the natural supply of amino acids and non-protein nitrogen present in milk is insufficient to support growth to high cell numbers. Similar to *L. lactis*, the cells rely on cell wall associated proteinases or use of protein hydrolysates such as EHC-K or whey hydrolysates as a source of amino acids, peptides, and oligopeptides. But unlike lactococci, *S. thermophilus* strains are weakly proteolytic. They require fewer amino acids than lactococci and lactobacilli. So far, for all the strains tested, only Gln and Glu along with sulfur- containing amino acids are considered to be essential amino acids. Also, the branched chain amino acid (BCAA) biosynthetic pathway is functional but insufficient to ensure optimal growth of *S. thermophilus* in the absence of BCAAs (Garault et al. 2000), which therefore need to be supplemented.

Protein hydrolysates in combination with yeast extract can provide an appropriate nitrogen source for optimal production of *S. thermophilus*. However, in some cases, additional criteria have to be considered when designing fermentation media for EPS producing strains. The production of EPS (capsules and/or extracellular EPS) can cause inefficient separation i.e., and have a negative effect on biomass yield. In *S. thermophilus* strain S2, that produces EPS capsules in milk, increased solids from 8% to 14% stimulated the production of larger capsules (Hassan et al. 2001). Additionally, increased lactose content (5%) and addition of whey protein concentrate or casein digest at 7 g/L in Elliker broth resulted in production of larger capsules.

Similar to lactobacilli, several studies investigated the effect of certain casein hydrolysates on growth and viability of *S. thermophilus* in milk. Addition of hydrolysates of casein and whey to milk enhanced the growth and acidification rate of *S. thermophilus* ST-7, thus reducing the fermentation time of yogurt (Lucas et al. 2004). Casein and whey peptones studied in this work were CH1, CH2, WPH1, WPH2, and WPH3 (Armor Proteines, Saint Brice en Cogles, France) which are used at levels ranging from 0.25 to 4 g/L.

Addition of 2% of acid casein hydrolysate and cysteine to milk demonstrated to improve the viability of the *S. thermophilus* WJ7 over 12 weeks, when tested in frozen dairy dessert (Ravula and Shah 1998).

Even though some of the protein hydrolysates may decrease the fermentation time, other additional criteria should be considered when selecting hydrolysates for direct application in fermented milks, since highly hydrolyzed protein hydrolysates have the propensity for bitterness.

Genus Bifidobacterium

Due to documented probiotic effect and health benefits, certain strains of *Bifidobacterium* sp. became important industrial strains. Well-documented strains are used both as components of starter cultures for fermented milks and as encapsulated freeze-dried material.

All bifidobacteria can utilize lactose, which allows them to grow in milk, although the growth is often weak due to low proteolytic activity (Klaver et al. 1993; Collins and Hall 1984). Bifidobacteria are a nutritionally heterogeneous group, but they are not as fastidious as other lactic acid bacteria. (Poupard et al. 1973; Kurmann 1988). Bifidobacteria normally do not require purines or pyrimidines (Kurmann and Rasic 1991) and are capable of degrading and utilizing proteins. *Bifidobacterium bifidum* and *Bifidobacterium adolescentis* produce intracellular and extracellular proteases. Most strains contain a leucine aminopeptidase, while a few have a valine aminopeptidase (Desjardins et al. 1990).

Most bifidobacteria are able to use ammonium salts as their only source of nitrogen. (Azaola et al. 1999), but supplementation of peptides and amino acids are considered a requirement for the economical production of these strains. Nitrogen requirements are known to be usually strain dependent, but the typical nitrogen sources are peptides/amino acids, cysteine, and ammonium salts. Bifidobacteria are rather demanding for growth factors. Of the vitamins, biotin and calcium pantothenate are the typical requirements (Kurmann and Rasic 1991). Several proteinaceous growth promoters, such as disulfide/sulfhydryl-containing peptides, lactoferrin with bound metals (Fe, Cu, Zn), α-lactalbumin, and ß-lactoglobulin are formed in milk (Petschow and Talbott 1991).

An unusual requirement for D-glucosamine is observed in *Bifidobacterium* sp. D-glucosamine is an essential part of the cell wall since it constitutes the backbone of the peptidoglycan units, N-acetylglucosamine and muramic acid (Poupard et al. 1973). This is of utmost importance since the composition and the strength of the cell wall determines the survival of bifidobacteria during downstream processes. Another nitrogen containing component that advantageously may be added is N-acetyl-glucosamine, which as mentioned above, is essential for cell wall production. Milk contains N-acetyl-glucosamine in the form of oligosaccharides and it is most often used as a glucosamine source (Exterkate and Veerkamp 1969).

As indicated above, *Bifidobacterium* sp. are only weakly proteolytic and proteolytic digest has to be supplemented into the growth medium. Small peptides are better amino acid source than free amino acids (Proulx et al. 1994). If the enzyme hydrolysis is the choice, then the selection of proteinase is very important for the value of a protein hydrolysate as a growth promoter. According to Proulx et al. (1994), peptides from trypsin-degraded casein have a better growth-promoting effect for *B. longum* and *B. infantis* than enzyme digests of Alcalase or chymotrypsin.

Growth and acid production by *B. lactis* strain (CSK, Leeuwarden, The Netherlands) was evaluated using ovine and caprine milk as media supplemented with milk protein hydrolysate (MPH) prepared by using commercial protease (*Asperigillus* sp. Protease 2A; Amano Pharmaceutical, Nagoya, Japan) (Gomes and Malcata 1998). Free amino acids and the amino acid fraction of MPH were also evaluated as nitrogen enrichment sources added at levels 25–50 mL/L. Addition of hydrolyzed milk into ovine and caprine milk had a positive effect on viable counts of *B. lactis* and it was also shown that MPH was a better growth promoter than free amino acids.

In order to optimize the biomass yield of *B. infantis,* a surface response methodology has been used with the often-used complex medium, TPYG (Azaola et al. 1999). The result indicated that increased yeast extract at the expense of the other nitrogen sources gives a more than two-fold higher yield. A similar conclusion was reached for *B. longum* with the same basic medium where an *E. coli* extract showed a significant growth promoting effect (Ibrahim and Bezkorovainy 1994).

For the cultivation of bifidobacteria, one has to keep in mind not only the complex nutritional requirements, but also the extreme sensitivity of these strains to oxygen. This trait presents a major challenge in keeping the culture stable in the dairy industry, but during production of starter cultures, the problem is usually overcome by adding substances that can maintain a low redox potential. Typically, cysteine, ascorbic acid or sodium sulphite is used for this purpose.

Genus Pediococcus

Pediococci are usually used as components of starter cultures for traditional fermented sausages. One has to keep in mind that unlike milk, meat is not pasteurized before inoculation with starter cultures and still contains large amounts of indigenous microflora. Therefore, meat starter cultures have to be adapted to meat proteins, active and competitive with as short a lag phase as possible. These specificities pose particular requirements in regards to the production of cultures, e.g., the design of fermentation media.

For the production of Pediococcus sp. glucose or sucrose are used as energy and carbon sources. Growth rate was shown to be slightly faster on glucose (Henrik Albahn, CH-internal communication). Although not essential, addition of acetate was shown to decrease the lag phase and stimulate the growth of the organism.

Pediococcus pentosaceus is the strain with the largest number of absolute amino acid requirements: Val, Ala, Met, Pro, Arg, Glu, Cys, Tyr, and His are essential, while other amino acids have a stimulatory effect (Henrik Moellgaard, CH-internal communication). As in *L. lactis* (Law et al. 1976), addition of Cys-hydrochloride is stimulatory but the stimulative effect of cysteine is mainly observed in large-scale fermentations where it is used as an oxygen scavenger.

Casein peptone (hydrolyzed casein), primatone (hydrolyzed meat proteins) and a yeast paste were tested as nitrogen sources for the production of *P. pentosaceus*. Omission of primatone and casein peptone and use of only yeast extract resulted in a prolongation of the fermentation time. Reduction of levels of either primatone, yeast extract and casein peptone to 50% of the initial concentration resulted in reduction of biomass yield by 20–50% (Benedikte Grenov, CH-internal communication). An initial nitrogen/carbon (N/C) ratio between 2 and 4 can be considered as optimal.

Effects of various dried and frozen meat stocks (1–5%) were also evaluated as a possible growth promoter for *Pediococcus acidilactici* PA-1 (Chr. Hansen's, Inc.). The performance of the meat protein was affected by the ability of the culture to hydrolyze these meat proteins. Dried beef stock seemed to be the best for growth of the strain.

Summary

Different protein hydrolysates have been successfully used for fermentations of LAB such as N-Z amine, Hy-Case, Hy-Soy, Edamin, and N-Z- Case (Misono et al. 1985; Molskness et al. 1973; Gonzalez 1984; Vedamuthu 1980; Kegel and Wallace 1989). A more economical alternative would be to use commercial proteolytic enzymes (e.g. Neutrase from Novozymes) to degrade cheaper protein sources e.g. vegetable proteins and soy powder (Proulx et al. 1994; Lucas et al. 2004; Gomes and Malcata 1998).

As was mentioned, the most preferred nitrogen source would contain (1) a combination of oligopeptides that would support the initial growth, (2) di- and tri-peptides and essential amino acids, all of which can be transported across the cell membrane and used for biomass synthesis and requiring the least amount of energy (ATP).

In order to fulfill the complex set of growth requirements, peptones are usually combined with yeast extracts, which are an excellent source of peptides, vitamins and minerals. If needed, the medium can be supplemented with additional amounts of specific compounds as growth promoters. The level of application of nitrogen sources usually ranges from 0.5% to 2.5% (w/v), and they have to be carefully balanced with the concentration of a carbon source (C/N balance). Increased concentrations of a nitrogen source in a medium can lead to substrate inhibition (Henrik Albahn 2004; personal communication).

Protein hydrolysates originate from different sources and they are available at different levels of purity, ranging from technical to refined grade, food to non-food grade and also available as Kosher. Keeping in mind that a wide range of hydrolysates are available in the market, it seems like formulating an optimal growth medium for a specific strain of LAB is a fairly easy task. One can always find a combination of hydrolysates which will fulfill the nutritional requirements for a strain/culture. However, there are additional criteria/limitations which should also be taken into consideration while designing an optimum growth medium.

In enzymatic and acid hydrolysis, certain amino acids e.g. cysteine (Cys) in casein hydrolysates and tryptophan (Trp) in acid-digested casein peptone can be denatured to different levels making it unavailable for bacterial utilization (Fig. 2). Such details are particularly important when re-designing a medium that originally contained for example, pancreatic digest of casein that contains high levels of free Trp.

In some cases meat or soy hydrolysates contain higher levels of certain peptides and amino acids (Table 4), which make them to be a more viable option than casein peptones. However, due to the regulatory requirements such as Kosher, GMO-free, and BSE-free, the raw materials for the production of dairy starter cultures are usually limited to dairy derived protein hydrolysates. Certain food grade meat and vegetable hydrolysates are usually used in the production of probiotics, meat cultures, bread and vegetable cultures. Additional requirements that fermentation specialists should be aware of are the selection of hydrolysates for production of probiotic nutritional products, where allergen-free products are usually preferred.

In this chapter, we focused only on the most important LAB species and strains and reviewed growth requirements of many single strains. Though, this information is limited to production of single strains, it can give valuable data on basic needs of the organism. Production of undefined or defined starter cultures composed of several species/strains is much more complex, having in mind that in order to maintain reproducible performance, the ratio between strains has to be preserved throughout the sub-cultivation and fermentation process by using the correctly balanced carbon and nitrogen sources in a fermentation growth medium.

To summarize, there is no universal formula, which can give us answers as how to design an optimal growth medium and which ingredients to use. Some strains that belong to the same species might have similar growth requirements, but this should not be taken as a hard and fast rule. Optimization of fermentation medium and process for each industrial strain/starter culture, requires extensive and thorough studies but the above information can be used as a guide to produce successful fermentations.

References

Azaola A, Bustamante P, Huerta S, Saucedo G, Gonzalez R, Ramos C et al (1999) Use of surface response methodology to describe biomass production of *Bifidobacterium infantis* in complex media. Biotech Tech 13(2):93–95

Bringel F, Hubert JC (2003) Extent of genetic lesions of the arginine and pyrimidine biosynthetic pathways in *Lactobacillus plantarum, L. paraplantarum, L. pentosus,* and *L. casei*: prevalence of CO(2)-dependent auxotrophs and characterization of deficient arg genes in *L. plantarum.* Appl Environ Microbiol 69(5):2674–2683

Champagne CP (1998) Production de ferments lactiques dans l'industrie latiere. Centre de Recherche et de development sur le aliments Agriculture et Agroalimentaire Canada, Quebec, Canada

Chervaux C, Ehrlich SD, Maguin E (2000) Physiological study of *Lactobacillus delbrueckii* subsp. *bulgaricus* Strains in a novel chemically defined medium. Appl Environ Microbiol 66(12):5306–5311

Collins EB, Hall BJ (1984) Growth of bifidobacteria in milk and preparation of *Bifidobacterium infantis* for a dietary adjunct. J Dairy Sci 67(7):1376–1380

Dave RI, Shah NP (1998) Ingredient supplementation effects on viability of probiotic bacteria in yogurt. J Dairy Sci 81(11):2804–2816

Desjardins M-L, Roy D, Goulet J (1990) Uncoupling of Growth and acids production in *Bifidobacterium ssp.* J Dairy Sci 73(11):1478–1484

Elli M, Zink R, Rytz A, Reniero R, Morelli L (2000) Iron requirement of *Lactobacillus* spp. in completely chemically defined growth media. J Appl Microbiol 88(4):695–703

Exterkate FA, Veerkamp JH (1969) Biochemical changes in *Bifidobacterium bifidum* var. *pennsylvanicus* after cell wall inhibition. I. Composition of lipids. Biochim Biophys Acta 176(1):65–77

Foucaud C, Hemme D, Desmazeaud M (2001) Peptide utilization by *Lactococcus lactis* and *Leuconostoc mesenteroides*. Lett Appl Microbiol 32(1):20–25

Garault P, Letort C, Juillard V, Monnet V (2000) Branched-chain amino acid biosynthesis is essential for optimal growth of *Streptococcus thermophilus* in milk. Appl Environ Microbiol 66(12):5128–5133

Gomes AMP, Malcata FX (1998) Use of small ruminants' milk supplemented with available nitrogen as growth media for Bifidobacterium lactis and *Lactobacillus acidophilus*. J Appl Microbiol 85(5):839–848

Gonzalez CF (1984) Preservation of foods with non-lactose fermenting *Streptococcus lactis* subspecies *diacetilactis*. US Patent 4,477,471

Hassan AN, Frank JF, Shalabi SI (2001) Factors affecting capsule size and production by lactic acid bacteria used as dairy starter cultures. Int J Food Microbiol 64(1–2):199–203

Hebert E, Raya RR, De Giori GS (2000) Nutritional requirements and nitrogen-dependent regulation of proteinase activity of *Lactobacillus helveticus* CRL 1062. Appl Environ Microbiol 66(12):5316–5321

Hellinick S, Richard J, Juillard V (1997) The effects of adding lactococcal proteinase on the growth rate of *Lactococcus lactis* in milk depend on the type of enzyme. Appl Environ Microbiol 63(6):2124–2130

Hoier E, Janzen T, Henriksen CM, Rattray F, Brockmann E, Johansen E (1999) The production, application and action of lactic cheese starter cultures. In: Law BA (ed) Technology of cheesemaking. CRC Press, Boca Raton, FL
Hsieh CM, Yang FC, Iannotti EL (1999) The effect of soy protein hydrolyzates on fermentation by *Lactobacillus amylovorus*. Process Biochem 34(2):173–179
Hugenholtz J, Dijkstra M, Veldkamp H (1987) Amino acid limited growth of starter cultures in milk. FEMS Microbiol Lett 45:191–198
Ibrahim S, Bezkorovainy A (1994) Growth-promoting factors for Bifidobacterium longum. J Food Sci 59(1):189–191
Juillard V, Le Bars D, Kunji ER, Konings WN, Gripon JC, Richard J (1995) Oligopeptides are the main source of nitrogen for *Lactococcus lactis* during growth in milk. Appl Environ Microbiol 61(8):3024–3030
Juillard V, Foucaud M, Desmazeaud M, Richard J (1996) Utilisation des sources d'azote du lait par *Lactococcus lactis*. Lait 76(1–2):13–24
Juillard V, Guillot A, Le Bars D, Gripon JC (1998) Specificity of milk peptide utilization by *Lactococcus lactis*. Appl Environ Microbiol 64(4):1230–1236
Kegel MA, Wallace DL (1989) Use of stabilizing agents in culture media for growing acid producing bacteria. US patent 4,806,479
Kenny O, Fitzgerald RJ, Cuinn GO, Beresford TP, Jordan K (2003) Growth phase and growth medium effects on the peptidase activities of *Lactobacillus helveticus*. Int Dairy J 13(7):509–516
Klaver FA, Kingma F, Veerkamp AH (1993) Growth and survival of bifidobacteria in milk. Netherlands Milk Dairy J 47(3–4):151–164
Konings WN, Poolman B, Driessen AJ (1989) Bioenergetics and solute transport in Lactococci. Crit Rev Microbiol 16(6):419–476
Kunji ERS, Hagting A, De Vries CJ, Juillard V, Haandrikman AJ, Poolman B, Konings NW (1995) Transport of B-Casein-derived peptides by the oligopeptide transport system is a crucial step in the proteolytic pathway of *Lactococcus lactis*. J Biol Chem 270(4):1569–1574
Kunji ER, Mierau I, Hagting A, Poolman B, Konings WN (1996) The proteolytic systems of lactic acid bacteria. Antonie Leeuwenhoek 70(2–4):187–221
Kurmann JA (1988) Starters with selected intestinal bacteria. Fermented milk: science and technology. IDF Bulletin No. 227, Brussels
Kurmann JA, Rasic JL (1991) The health potential of products containing bifidobacteria. In: Robinson RK (ed) Therapeutic properties of fermented milks. London, Elsevier, pp 117–158
Law BA, Sezgin E, Sharpe ME (1976) Amino acid nutrition of some commercial cheese starters in relation to their growth in peptone-supplemented whey media. J Dairy Res 43(2):291–300
Loader NM, Lindner N, Pasupuleti VK (1999) Proteolytic system of lactic acid bacteria and nutrients. In: Nagodawithana TW, Reed G (ed) Nutritional Requirements of commercially important microorganisms. Esteekay Associates, Inc. Milwaukee, WI.
Lucas A, Sodini I, Monnet C, Jolivet P, Corrieau G (2004) Probiotic cell counts and acidification in fermented milks supplemented with milk protein hydrolysates. Int Dairy J 14(1):47–53
Mierau I, Kunji ER, Venema G, Kok J (1997) Casein and peptide degradation in lactic acid bacteria. Biotechnol Genet Eng Rev 14:279–301
Misono H, Norihiko G, Nagasaki S (1985) Purification, crystallization and properties of $NADP^+$ - Specific glutamate dehydrogenase from *Lactobacillus fermentum*. Agric Biol Chem 49(1):117–123
Molskness TA, Lee DR, Sandine WE, Elliker PR (1973) b-D-Phosphogalactohydrolase of Lactic Streptococci. Appl Micriobiol 25(3):373–380
Morel F, Frot-Coutaz J, Aubel D, Portalier R, Atlan D (1999) Characterization of a prolidase from *Lactobacillus delbrueckii* subsp. *bulgaricus* CNRZ 397 with an unusual regulation of biosynthesis. Microbiology 145(2):437–446
Morishita T, Fucada T, Shirota M, Yura T (1974) Genetic basis of nutritional requirements in *Lactobacillus casei*. J Bacteriol 120(1):1078–1084

Morishita T, Deguchi Y, Yajima M, Sakurai T, Yura T (1981) Multiple nutritional requirements of lactobacilli: genetic lesions affecting amino acid biosynthetic pathways. J Bacteriol 148(1):64–71

Norris RF, De Spin M, Zilliken FW, Harvey TS, Gyorgy P (1954) Occurrence of mucoid variants of *Lactobacillus bifidus*; demonstration of extracellular and intracellular polysaccharide. J Bacteriol 67(2):159–166

Novak L, Cocaign-Bousquet M, Lindley ND, Loubiere P (1997) Metabolism and energetic of *Lactococcus lactis* during growth in complex or synthetic media. Appl Environ Microbiol 63(7):2665–2670

Petersen BL, Dave RI, McMahon DJ, Oberg CJ, Broadbent JR (2000) Influence of capsular and ropy exopolysaccharide-producing *Streptococcus thermophilus* on Mozzarella cheese and cheese whey. J Dairy Sci 83(9):1952–1956

Petschow BW, Talbott RD (1991) Response of bifidobacterium species to growth promoters in human and cow milk. Pediatr Res 29(2):208–213

Poolman B, Konings WN (1988) Relation of growth of *Streptococcus lactis* and *Streptococcus cremoris* to amino acid transport. J Bacteriol 170(2):700–707

Porubcan RS, Sellars RL (1979) Lactic starter culture concentrates. In: Peppler HJ, Perlman D (eds) Microbial technology, vol 1, Microbial processes. Academic, New York, pp 59–92

Poupard J, Husain I, Norris RF (1973) Biology of the bifidobacteria. Bacteriol Rev 37(2):136–165

Proulx M, Ward P, Gauthier SF, Roy D (1994) Comparison of bifidobacterial growth-promoting activity of ultrafiltered casein hydrolyzate fractions. Lait 74(2):139–152

Ravula RR, Shah NP (1998) Effect of acid casein hydrolysate and cysteine on the viability of yogurt and probiotic bacteria in fermented frozen dairy desserts. Austr J Dairy Technol 53(3):175–179

Smid EJ, Konings WN (1990) Relationship between utilization of proline and proline-containing peptides and growth of *Lactococcus lactis*. J Bacteriol 172(9):5286–5292

Summer R (1996) Yeast extracts: production, properties and components. Paper presented at the 9th international symposium on yeasts, Sydney, August 1996

ten Otto R, Brink B, Veldkamp H, Konings WN (1983) The relationship between growth rate and electrochemical proton gradient of *Streptococcus cremoris*. FEMS Microbiol Lett 16:69–74

Vedamuthu ER (1980) Method for diacetyl flavor and aroma development in creamed cottage cheese. US Patent 4,191,782

Chapter 7
Protein Hydrolysates from Non-bovine and Plant Sources Replaces Tryptone in Microbiological Media

Yamini Ranganathan, Shifa Patel, Vijai K. Pasupuleti, and R. Meganathan

Abstract Tryptone (pancreatic digest of casein) is a common ingredient in laboratory and fermentation media for growing wild-type and genetically modified micro-organisms. Many of the commercially manufactured products such as human growth hormone, antibiotics, insulin, etc. are produced by recombinant strains grown on materials derived from bovine sources. With the emergence of Bovine Spongiform Encephalopathy (BSE) and the consequent increase in Food and Drug Administration (FDA) regulations, elimination of materials of bovine origin from fermentation media is of paramount importance. To achieve this objective, a number of protein hydrolysates derived from non-bovine animal and plant sources were evaluated. Tryptone in Luria-Bertani (LB) broth was replaced with an equal quantity of alternate protein hydrolysates. Four of the six hydrolysates (one animal and three from plants) were found to efficiently replace the tryptone present in LB-medium as measured by growth rate and growth yield of a recombinant *Escherichia coli* strain. In addition, we have determined plasmid stability, inducibility and activity of the plasmid encoded β-galactosidase in the recombinant strain grown in the presence of various protein hydrolysates.

Keywords Plant protein hydrolysates • Tryptone • Plasmid stability • β-galactosidase

---

Y. Ranganathan, S. Patel*, and R. Meganathan (✉)
Department of Biological Sciences, Northern Illinois University, DeKalb, IL 60115, USA
*Undergraduate Research Participant
e-mail: rmeganathan@niu.edu

V.K. Pasupuleti
SAI International, Geneva, IL 60134, USA

V.K. Pasupuleti and A.L. Demain (eds.), *Protein Hydrolysates in Biotechnology*,
DOI 10.1007/978-1-4020-6674-0_7,

Introduction

The inclusion of animal-derived materials in microbiological media has a long history. The advent of recombinant DNA technology and its incorporation into the manufacture and production of bio-pharmaceutical products has resulted in the utilization of vast quantities of media containing animal derived components. A short list of products produced by utilizing recombinant DNA technology by a variety of bacteria grown in media containing animal-derived components are presented in Table 1.

However, with the discovery of the wide spread outbreak of prion disease, Bovine Spongiform Encephalopathy (BSE, Mad cow disease) in cattle in Europe and the isolated cases reported from Japan, Canada and the United States, there is concern about transmission of the disease to humans from contaminated materials (Lister and Becker 2007; Prusiner 2004; Brown et al. 2001). Transmission of BSE due to consumption of infected animals has been established. The Human form of BSE is referred to as variant Creutzfeld – Jakob disease (vCJD).

The well established processes employed in the manufacture and preparation of laboratory and industrial media are sufficient to prevent the transmission of animal diseases via culture media. Because of its unusual and unique properties, BSE prion can withstand many of these treatments (Appel et al. 2001). It has been shown that prions can withstand autoclaving for 30 h, resist microwaves, ultrasound and ionizing radiation. In addition, they are resistant to such extremely harsh materials as acids, bases, alkylating agents, detergents, and organic solvents (Narang 2002; Prusiner 1997).

The problem of BSE contamination is further compounded by the fact that there is no sensitive assay for the detection of such agents in fermentation media (Behizad and Curling 2000). Recognizing the potential hazard to humans and animals due to contaminated pharmaceutical and biological materials, the World Health Organization (WHO) and the Food and Drug Administration (FDA) recommend that the manufacturers of regulated products do not use materials from cattle born, raised or slaughtered in countries where BSE exists (Lister and Becker 2007). Two other alternatives that have been suggested and being pursued are use of materials derived from: (1) non-bovine animals and (2) plant-derived materials. In this report, comparative studies on replacing tryptone in Luria-Bertani (LB) medium by materials derived from non-bovine sources and plant products for supporting growth, plasmid maintenance, and expression of β-galactosidase activity are presented.

Materials and Methods

Bacterial Strain

The organism used in this study was *Escherichia coli* ATCC 39114 carrying plasmid pOP(UV-5)-3. The plasmid contains the *lacZ* gene and expresses β-galactosidase to a very high level (up to 15% of total cell protein).

Table 1 Media containing animal products used for growth of various recombinant bacteria for the production of nutritionals and bio-pharmaceuticals

Organism	Product	Media used	Reference
Bacillus subtilis	Fibrinolytic enzymes	LB	Wang et al. 2006
Corynebacterium ammoniagenes	Riboflavin	Seed medium (contains peptone)	Koizumi et al. 2000
Escherichia coli	Interleukin-11	M9Ca (contains tryptone)	Zhao et al. 2005
Escherichia coli	Vanillin	LB	Torre et al. 2004
Escherichia coli	TGF-a-PE40 growth factor	Complex medium M101 (contains peptone)	Lee et al. 1997
Escherichia coli	Human epidermal growth factor (hEGF)	YT and MMBL media (contains tryptone)	Sivakesava et al. 1999
Escherichia coli	Human granulocyte colony-stimulating factor (hG-CSF)	LB	Fallah et al. 2003
Escherichia coli	Resistin	YT medium	Juan et al. 2003
Escherichia coli	*Treponema pallidum* Membrane Protein A	LB and M9 medium with 0.5% tryptone	Rivera et al. 1999
Escherichia coli	Abl Protein Tyrosine Kinase	LB	Songyang et al. 1995
Escherichia coli	Human Chymase	LB	Takai et al. 2000
Escherichia coli	Gonadotropin-releasing hormone	LB	Xu et al. 2006
Escherichia coli	Erythromycin C	LB	Peiru et al. 2005
Escherichia coli	Ansamycin polyketide precursor	LB	Watanabe et al. 2003
Escherichia coli	Aminoglycoside Antibiotic 3-Acetyltransferase-IIIb	LB	Owston and Serpersu 2002
Escherichia coli	Human Parathyroid Hormone	LB	Wingender et al. 1989
Escherichia coli	Recombinant Human Parathyroid Hormone 1–34	Terrific broth (contains tryptone)	Suzuki et al. 1998
Escherichia coli	Human Cardiac-Specific Homeobox Protein	LB	Zhao et al. 2000
Escherichia coli	Human growth hormone	LB	Tabandeh et al. 2004
Escherichia coli	Human Tissue Plasminogen Activator	Super broth	Manosroi et al. 2001
Pseudomonas sp. No. 57-250	Micacocidin A, B and C	LB	Kobayashi et al. 1998
Streptomyces lividans	Cephamycin	YEME (contains peptone)	Coque et al. 1996

Media

For routine growth of the culture LB agar and LB broth were used. LB medium contained tryptone 10 g, yeast extract 5 g and sodium chloride 10 g/L of distilled water. The tryptone and yeast extract were obtained from Fisher Scientific.

The culture was stored in glycerated LB broth at –80°C. Tetracycline was used at a concentration of 20 μg/mL. IPTG (Isopropyl-β-D-thiogalactopyranoside) was used at a concentration of 0.5 mM.

The six protein hydrolysates replacing tryptone in LB medium were: Medium 1, enzymatic hydrolysate of casein (bovine); Medium 2, Enhancetone (enzymatic digest of gelatin from porcine and yeast); Medium 3, soy protein (enzymatically hydrolyzed grits); Medium 4, wheat protein (enzymatically hydrolyzed wheat gluten); Medium 5, pea protein (enzymatic hydrolysate); and Medium 6, enzymatic hydrolysate of whey protein and casein (bovine). All the six protein hydrolysates used in this study were supplied by DMV international, N.Y. The concentration of protein hydrolysates used was 10 g/L (w/v).

Growth Studies

For growth studies, seed cultures were grown aerobically in 5 mL of LB broth, contained in test tubes (16 by 150 mm), at 37°C. A 1% inoculum was used to inoculate 20 mL of medium contained in 500 mL sidearm flasks and grown with shaking at 250 rpm at 37°C. Growth was monitored in a Klett colorimeter with a red filter (wavelength, 640–700 nm).

β-Galactosidase Assay

For the assay of β-galactosidase, a single colony of the culture was inoculated as a starter culture into 5 mL of the appropriate medium containing tetracycline (20 μg/mL), contained in test tubes (16 by 150 mm), and incubated with shaking at 250 rpm at 37°C. After 8–12 h, a 1% inoculum was transferred into 5 mL of fresh medium of the same composition and grown aerobically with shaking overnight. After overnight growth, a 1% inoculum was transferred into 20 mL of fresh medium contained in a 125 mL Erlymeyer flask and growth continued until a cell density of 50 and 100 Klett units respectively was reached. β-galactosidase was induced by the addition of IPTG at a final concentration of 0.5 mM. The culture was induced for 30 min at 37°C with shaking at 250 rpm. Cells grown as described above were assayed for β-galactosidase activity and the units calculated as described by Miller (1992).

Plasmid Stability

To determine the plasmid stability in various media, a single colony of the culture was inoculated into 5 mL of LB broth, contained in test tubes (16 by 150 mm), and incubated with shaking at 250 rpm at 37°C. After overnight growth, a 1% inoculum was transferred into 20 mL of various media (medium 1–medium 6 and LB) contained in 125 mL Erlenmeyer flasks and grown with shaking at 250 rpm at 37°C. The culture was grown up to a Klett reading of 200. Samples (100 μL) were withdrawn, diluted and plated on various media in the presence and absence of tetracycline. The plates were incubated at 37°C for 12 h and colonies were counted.

Results and Discussion

Properties of the Protein Hydrolysates

As mentioned in the introduction, six different protein hydrolysates, three from animal sources (two bovine and one porcine) and three from plant sources were tested for their ability to replace tryptone in LB medium. For comparison LB medium containing tryptone (Fisher Scientific) was used as a control. The chemical properties of the hydrolysates such as protein content, total nitrogen, amino nitrogen, pH and moisture content are shown in Table 2.

As seen from the table, the protein content of both animal-derived and plant-derived materials is about the same. The amino nitrogen content of the plant proteins was lower than that of the bovine derived materials casein (medium 1) and whey protein concentrate and casein (medium 6).

It is worth noting that among the plant derived materials, pea protein (medium 5) had as much amino nitrogen as porcine gelatin and yeast (medium 2). The total nitrogen was approximately the same in all media. The pH of a 1% solution of two bovine and the three plant materials was around 7.0 while porcine-derived material had a pH of <5.0.

The compositions of amino acids present in all six media are shown in Table 3. As seen from the table all the amino acids were more or less present in all of the media studied.

Growth of *E. coli* in Various Media

The ability of various protein hydrolysates to replace tryptone in LB medium was determined by the measurement of growth. All the six media and LB as control were inoculated as described in Methods and growth was monitored. The results are presented in Fig. 1.

Table 2 Chemical composition of the various media

Components	Various media						
	Tryptone as in LB	Medium 1	Medium 2	Medium 3	Medium 4	Medium 5	Medium 6
Protein (%)	NA	86.1	82.3	78.0	79.0	84.0	80.4
Amino nitrogen (AN) (%)	4.7	4.3	2.9	2.0	1.0	3.0	5.4
Total nitrogen (TN) (%)	13.0	13.5	12.8	12.0	14.0	13.0	12.6
AN/TN	36.1	31.9	22.7	19.0	17.0	20.0	42.9
Ash (%)	8.1	<6.0	<5.0	<10.0	<10.0	<10.0	<6.0
Moisture (%)	NA	<5.0	6.9	<6.0	<6.0	<6.0	<6.0
pH of 1% solution	6.7	7.0	<5.0	7.6	7.1	7.4	7.1

NA = Not available

Table 3 Percentage amino acid composition of the various media

	Various media						
Amino acids	Tryptone as in LB	Medium 1	Medium 2	Medium 3	Medium 4	Medium 5	Medium 6
Alanine	NA	2.9	6.2	3.4	2.1	3.8	3
Arginine	NA	3.3	5.5	5.8	2.5	7.6	2.3
Aspartic acid	NA	6.5	6.7	9.1	2.2	10.2	9
Cystine	NA	0.3	0.2	1.2	1.9	0.8	2.8
Glutamic acid	NA	20.2	9.4	16.1	36.5	15.8	16.3
Glycine	NA	1.8	14.7	0.2	3.1	3.5	1.4
Histidine	NA	2.3	0.7	1.9	1.5	1.8	1.4
Isoleucine	NA	5.1	1.4	3.3	2.9	4.1	4.6
Leucine	NA	8.6	2.6	5.4	5.5	7	7.5
Lysine	NA	7.4	3.3	5.1	1.1	6.1	6.6
Methionine	NA	2.6	0.7	1	1.2	0.8	1.7
Phenylalanine	NA	4.3	1.9	3.2	4.8	4.5	2.8
Proline	NA	9.6	8.4	3.8	12.5	4.1	6.6
Serine	NA	4.8	2.6	3.8	4.1	4.3	3.7
Threonine	NA	3.8	1.7	3.1	2.1	3	3.8
Tryptophan	NA	0.5	0.1	0.5	0.3	0.3	0.9
Tyrosine	NA	1.9	0.8	2.4	1.4	3.1	0.9
Valine	NA	6.2	2.2	2.5	3.1	4.3	4.8

NA – not available. Fisher Scientific Company failed to provide the composition in spite of repeated requests

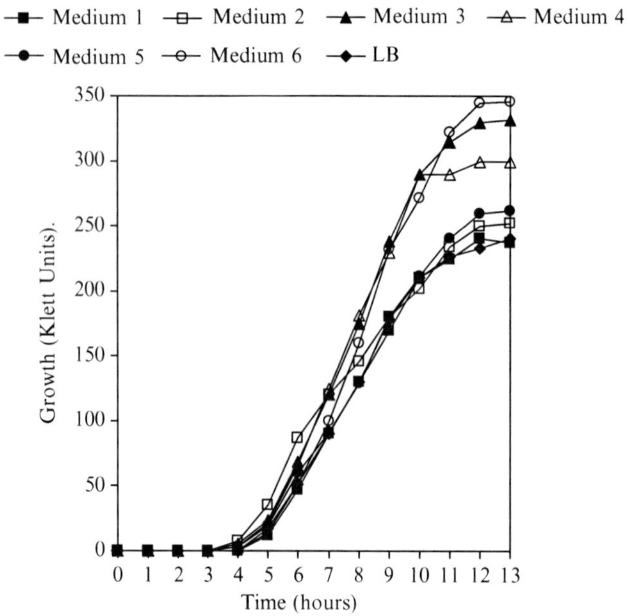

Fig. 1 Growth curve of *E. coli* in various media

It can be seen from the figure that the various media are equal or better than LB in supporting growth. The casein (medium 1) supported identical growth to that of the control (LB) and the pea protein (medium 5) and porcine gelatin and yeast (medium 2) were equal. However, soy (medium 3), wheat (medium 4) and whey protein concentrate and casein (medium 6) were somewhat better than the other four media.

Expression of Plasmid Borne β-Galactosidase

The ability of various media to support induction of β-galactosidase was determined. The results are shown in Fig. 2. As seen from the figure, all the different media supported expression

Stability of Plasmid

In order to determine the stability of the plasmid, various media were inoculated with the culture and grown in the absence of tetracycline. When the cultures reached a Klett reading of about 200, they were diluted and plated on media with

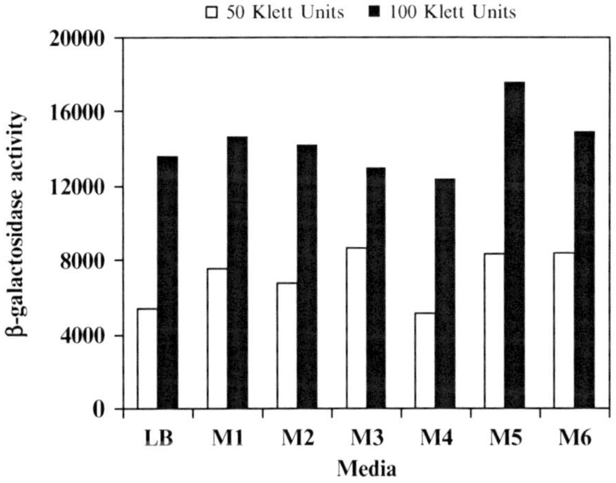

Fig. 2 β-Galactosidase activity of *E. coli* in various media

Table 4 Stability of the plasmid pOP(UV-5)-3 in various media

Media	Total count	Tet^r colonies	Tet^s colonies	Plasmid loss (%)
LB	80	58	22	28
Medium 1	201	144	57	28
Medium 2	271	206	65	24
Medium 3	116	86	30	26
Medium 4	173	122	51	29
Medium 5	121	87	34	28
Medium 6	213	161	52	24

and without tetracycline. After growth, the total count and the Tet^r colonies were determined (Table 4). It can be seen from the table, that the percent loss of the plasmid was approximately the same (25–30%) irrespective of the media employed.

Conclusions

The spread of BSE in England is mainly due to feeding cattle with processed slaughterhouse waste. In short vegetarian cattle were fed a non-vegetarian diet, which resulted in the spread of BSE. This study demonstrates that vegetable-derived protein hydrolysates in combination with yeast extract are as effective for microbial growth as animal-derived protein hydrolysates. Thus, growing microbes on a purely vegetarian diet for preparation of products for human use should keep BSE away from humans!

References

Appel TR, Wolff M, von Rheinbaben F, Heinzel M, Riesner D (2001) Heat stability of prion rods and recombinant prion protein in water, lipid and lipid–water mixtures. J Gen Virol 82:465–473

Behizad M, Curling JM (2000) Comparing the safety of synthetic and biological ligands used for purification of therapeutic proteins. Biopharm 13:42–46

Brown P, Will RG, Bradley R, Asher DM, Detwiler L (2001) Bovine Spongiform Encephalopathy and variant Creutzfeldt-Jakob disease: background, evolution, and current concerns. Emerg Infect Dis 7:6–16

Coque JJR, De La Fuente JL, Liras P, Martín JF (1996) Overexpression of the *Nocardia lactamdurans* α-aminoadipyl-cysteinyl-valine synthetase in *Streptomyces lividans*. The purified multienzyme ses cystathionine and 6-Oxopiperidine 2-Carboxylate as substrates for synthesis of the tripeptide. Eur J Biochem 242:264–270

Fallah MJ, Akbari B, Saeedinia AR, Karimi M, Vaez M, Zeinoddini M, Soleimani M, Maghsoudi N (2003) Overexpression of recombinant human granulocyte colony-stimulating factor in *E. coli*. Ir J Med Sci 28:131–134

Juan CC, Kan LS, Huang CC, Chen SS, Ho LT, Au LC (2003) Production and characterization of bioactive recombinant resistin in *Escherichia coli*. J Biotechnol 103:113–117

Kobayashi S, Nakai H, Ikenishi Y, Sun WY, Ozaki M, Hayase Y, Takeda R (1998) Micacocidin A, B and C, novel antimycoplasma agents from *Pseudomonas sp*. II. Structure elucidation. J Antibiot (Tokyo) 51:328–332

Koizumi S, Yonetani Y, Maruyama A, Teshiba S (2000) Production of riboflavin by metabolically engineered *Corynebacterium ammoniagenes*. Appl Microbiol Biotechnol 53:674–679

Lee C, Sun WJ, Burgess BW, Junker BH, Reddy J, Buckland BC, Greasham RL (1997) Process optimization for large-scale production of TGF-a-PE40 in recombinant *Escherichia coli*: effect of medium composition and induction timing on protein expression. J Ind Microbiol Biotechnol 18:260–266

Lister SA, Becker GS (2007) Bovine Spongiform Encephalopathy (BSE, or "Mad Cow Disease"): current and proposed safeguards. CRS Report RL32199. Congressional Report Service, Washington DC

Manosroi J, Tayapiwatana C, Götz F, Werner RG, Manosroi A (2001) Secretion of active recombinant Human tissue plasminogen activator derivatives in *Escherichia coli*. Appl Environ Microbiol 67:2657–2664

Miller JH (1992) A short course in bacterial genetics. Laboratory manual. Cold Spring Harbor Laboratory Press, New York

Narang H (2002) A critical review of the nature of the spongiform encephalopathy agent: protein theory versus virus theory. Exp Biol Med 227:4–19

Owston MA, Serpersu EH (2002) Cloning, overexpression, and purification of aminoglycoside antibiotic 3-acetyltransferase-IIIb: conformational studies with bound substrates. Biochemistry 41:10764–10770

Peiru S, Menzella HG, Rodriguez E, Carney J, Gramajo H (2005) Production of the potent antibacterial polyketide erythromycin C in *Escherichia coli*. Appl Environ Microbiol 71:2539–2547

Prusiner SB (1997) Prion diseases and the BSE crisis. Science 278:245–251

Prusiner SB (2004) Detecting Mad Cow Disease. Sci Am 291:86–93

Rivera JM, Domínguez MC, Ponce M, Narciandi RE (1999) Production of recombinant *Treponema pallidum* membrane protein A in *Escherichia coli* using fed-batch fermentation. Biotecnologia Apl 16:145–148

Sivakesava S, Xu ZN, Chen YH, Hackett J, Huang RC, Lam E, Lam TL, Siu KL, Wong RSC, Wong WKR (1999) Production of excreted human epidermal growth factor (hEGF) by an efficient recombinant *Escherichia coli* system. Process Biochem 34:893–900

Songyang Z, Carraway KL, Eck MJ, Harrison SC, Feldman RA, Mohammadi M, Schlessinger J, Hubbard SR, Smith DP, Eng C, Lorenzo MJ, Ponder BAJ, Mayer BJ, Cantley LC (1995) Catalytic specificity of protein-tyrosine kinases is critical for selective signaling. Nature 373:536–539

Suzuki Y, Yabuta M, Ohsuye K (1998) High-level production of recombinant human parathyroid hormone 1–34. Appl Environ Microbiol 64:526–529

Tabandeh F, Shojaosadati SA, Zomorodipour A, Khodabandeh M, Sanati MH, Yakhchali B (2004) Heat-induced production of human growth hormone by high cell density cultivation of recombinant *Escherichia coli*. Biotechnol Lett 26:245–250

Takai S, Sumi S, Aoike M, Sakaguchi M, Itoh Y, Jin D, Matsumura E, Miyazaki M (2000) Characterization of recombinant human chymase expressed in *Escherichia coli*. Jpn J Pharmacol 82:144–149

Torre P, De Faveri D, Perego P, Converti A, Barghini P, Ruzzi M, Faria FP (2004) Selection of co-substrate and aeration conditions for vanillin production by *Escherichia coli* JM109/pBB1. Food Technol Biotech 42:193–196

Wang CT, Ji BP, Li B, Nout R, Li PL, Ji H, Chen LJ (2006) Purification and characterization of a fibrinolytic enzyme of *Bacillus subtilis* DC33, isolated from Chinese traditional Douchi. J Ind Microbiol Biotechnol 33:750–758

Watanabe K, Rude MA, Walsh CT, Khosla C (2003) Engineered biosynthesis of an ansamycin polyketide precursor in *Escherichia coli*. Proc Natl Acad Sci USA 100:9774–9778

Wingender E, Bercz G, Blocker H, Frank R, Mayer H (1989) Expression of human parathyroid hormone in *Escherichia coli*. J Biol Chem 264:4367–4373

Xu J, Li W, Wu J, Zhang Y, Zhu Z, Liu J, Hu Z (2006) Stability of plasmid and expression of a recombinant gonadotropin-releasing hormone (GnRH) vaccine in *Escherichia coli*. Appl Microbiol Biotechnol 73:780–788

Zhao JH, Xu Z, Hua ZC (2000) Expression of human cardiac-specific homeobox protein in *Escherichia coli*. Protein Expr Purif 18:316–319

Zhao K, Gan Y, Zhang ZJ (2005) A study of fermentation for human Interleukin-11 with recombinant *E. coli*. Int J Chemical React Eng 3:A49

Chapter 8
The Use of Protein Hydrolysates for Weed Control

Nick Christians, Dianna Liu, and Jay Bryan Unruh

Abstract Corn gluten meal, the protein fraction of corn (*Zea mays* L.) grain, is commercially used as a natural weed control agent and nitrogen source in horticultural crops and in the turf and ornamental markets. Corn gluten hydrolysate, a water soluble form of gluten meal, has also been proposed for the same purpose, although it could be sprayed on the soil rather than applied in the granular form. Five depeptides, glutaminyl-glutamine (Gln-Gln), glycinyl-alanine (Gly-Ala), alanyl-glutamine (Ala-Glu), alanyl-asparagine (Ala-Asp), and alaninyl-alanine (Ala-Ala) and a pentapeptide leucine-serine-proline-alanine-glutamine (Leu-Ser-Pro-Ala-Gln) were identified as the active components of the hydrolysate. Microscopic analysis revealed that Ala-Ala acted on some metabolic process rather than directly on the mitotic apparatus. Similar to the chloracetamides and sulfonyl-urea hebicides, Ala-Ala inhibits cell division rather than disrupting of cell division processes. Cellular ultrastructure changes caused by exposure to Ala-Ala implicate Ala-Ala as having membrane-disrupting characteristics similar to several synthetic herbicides. The potential use of the hydrolysate and the peptides as weed controls is discussed.

Keywords Dipeptides • Natural weed control • Natural herbicide

N. Christians (✉)
Iowa State University, Ames, IA, USA
e-mail: nchris@iastate.edu

D. Liu
Innovative Food Technologies Agri-Science Queensland
A service of the Department of Employment, Economic Development and Innovation
Brisbane, Queensland, Australia

J.B. Unruh
Department of Environmental Horticulture, West Florida Research and Education Center,
IFAS, University of Florida, FL 32565, USA

V.K. Pasupuleti and A.L. Demain (eds.), *Protein Hydrolysates in Biotechnology*,
DOI 10.1007/978-1-4020-6674-0_8,

Introduction

Public concern over the use of synthetic pesticides in recent years has led to a search for natural products that can be used in their place. While considerable progress has been made in the use of natural materials and alternative methods for insect control, substitutes for synthetic weed controls have been limited. One commercially-available natural preemergence herbicide presently available in the US and Canada is composed of corn gluten meal, the protein fraction of corn grain. The material is generally applied in a dry, granulated form to the soil. It is also available in a light powdery form which is not soluble in water and must be spread by hand.

Hydrolyzed proteins of corn, soybean (*Glycine max*), and wheat (*Triticum* spp.), which are water soluble and sprayable, have also been developed as natural preemergence herbicides (Christians et al. 1994a). Several peptides responsible for the biological activity of the grain proteins have been identified (Christians et al. 1994b).

Corn Gluten Meal

In the mid 1980s, a research project using corn meal as a growing medium for a fungal organism led to an observation that corn grain contains compounds that inhibit the root formation of germinating seeds (Christians 1993; Liu and Christians 1997a). Further work demonstrated that the root inhibiting compounds were concentrated in the protein fraction of corn, the corn gluten meal (CGM).

Wet milling of corn grain is used to separate the starch, corn oil, defatted corn germ, hulls, steep liquor, and protein (gluten) from the grain. The gluten is separated from the starch stream by centrifugation. The resulting material is a yellow slurry that contains 15–20% solids. The slurry is dried to produce the CGM, which is a major protein source in livestock and pet foods (Christians et al. 1994a).

The CGM contains 10% nitrogen by weight and makes an excellent fertilizer for plants that have a well-established root system. It was later demonstrated that CGM could be used as a natural 'weed and feed' material in lawns and other horticultural areas (Christians 1993). The material acts as a preemergence only and has no postemergence effects on weeds that are already established. The controlled release of nitrogen from the CGM is similar to that of other natural nitrogen sources used in the turf and ornamental industry. US Patent #5,030,268 titled "Preemergence weed control using corn gluten meal" was issued in 1991 and was later reissued with broader claims under #Re. 34,594 (Christians 1991). Corn gluten meal is presently marketed by several companies in the US and Canada under a wide variety of commercial names (www.gluten.iastate.edu).

Grain Protein Hydrolysates

The next step in the research was to try to isolate the chemical or chemicals in the CGM that were responsible for the biological activity. To do this, graduate student Dianna Liu needed a water soluble form of corn gluten meal for use in a high performance liquid chromatograph (HPLC). We screened more than 50 products derived from CGM. We found that material developed by Grain Processing Corporation of Muscatine, Iowa called corn gluten hydrolysate provided a concentrated source of the active components in a water-soluble form that could be further studied to identify the chemistry of the compound or compounds responsible for the root-inhibiting activity (Liu and Christians 1994, 1997b; Liu et al. 1994).

The basic description of the corn gluten hydrolysate as listed in patent #5,290,749 is as follows:

> To prepare corn gluten hydrolysates, the liquid corn gluten (15–20% solids) is preferably diluted with water to a solids concentration of about 5 to 20% and the pH adjusted to about 6.0 to 8.0, preferably to about pH 6.5. The appropriate amylase is added [0.1 to 1.0% dry basis (DB)] and the slurry jet cooked at 280° to 340°F., preferably 320°F. for 3–4 minutes. The cooked slurry is then adjusted to about pH 4 to 5, cooled to 140°F. and, optionally, a saccharifying amylase (glucoamylase) is added (0.01 to 0.1% DB) and the slurry maintained at 140°F. for 8–18 hours, preferably about 12 h. The slurry is then filtered and washed and the filtrate and washings discarded. The filter cake is reslurried in water to 5 to 20% solids (preferably about 10%) and adjusted to pH 7.5 to 9 with $Ca(OH)_2$. An alkaline protease is then added (0.1% to 1% DB) and the slurry is maintained at 50° to 60°C. for 2 to 6 hours, or until the pH remains constant. The slurry is then adjusted to pH 6.0 to 6.8 (preferably pH 6.2), the precipitated $Ca(PO_4)_2$ and any insoluble residues are removed by filtration. The clear filtrate is then dried in an appropriate manner (i.e., spray drying, drum drying, etc.) to yield a dry solid product having greater than about 80–90% protein (Kjeldahl nitrogen), and which is essentially water-soluble at 10 wt--% concentration. On a dry basis, the corn gluten hydrolysate will have a nitrogen content of at least about 8%, i.e., about 8–11.2%, most preferably at least about 14.4%. (Christians et al. 1994a)

The patent goes on to describe the preparation of soy and wheat hydrolysates and to present alternative procedures for making the hydrolysates. It also presents data on the biological activity of the grain hydrolysates on a variety of plant species.

In later work, the biological activity of the enzymatically hydrolyzed corn gluten meal treated with either fungal or bacterial proteinases was compared against various samples derived from corn, soybean, and wheat (Liu et al. 1994). Separate tests demonstrated that the proteinases used in the hydrolysis process had no effect on rooting of the test species. The studies were conducted in the growth chamber in petri dishes and on soil in the greenhouse. The target plant species were smooth crabgrass (*Digitaria ischaemum* Schreb.), creeping bentgrass (*Agrostis palustris* Huds.), and perennial ryegrass (*Lolium perenne* L.). The corn gluten hydrolysate prepared with the bacterial proteinase was the most effective material at inhibiting root formation in these species. The corn gluten hydrolysate inhibited root formation of creeping bentgrass at application levels of 0.118 g/dm^2 and crabgrass at 0.236 g/dm^2 in Petri dishes. In soil, complete inhibition of seedling establishment was achieved at 4.69 g/dm^2.

Peptides Isolated from Hydrolysates

In further studies, we subjected an aqueous solution of corn gluten hydrolysate described above to column chromatography (gel filtration). Bioassays were conducted on the eluate to identify biologically active fractions capable of inhibiting root formation of perennial ryegrass. The active fractions were isolated and collected by using HPLC equipped with a reversed-phase column. The bioactive fractions were further purified and sequenced to determine the peptidyl components by the Iowa State University protein facility. The naturally occurring peptides identified in this procedure were resynthesized and the biological activity of the synthetically produced peptides was demonstrated (Christians et al. 1994b; Liu and Christians 1994).

We identified five individual dipeptides that were responsible for the root inhibition. They included glutaminyl-glutamine (Glu-Glu), glycinyl-alanine (Gly-Ala), alaninyl-glutamine (Ala-Glu), alanyl-asparagine (Ala-Asp), and alanyl-alanine (Ala-Ala). The idea of using these naturally occurring dipeptides as substitutes for synthetic preemergence herbicides was submitted to the US Patent office in 1993 and the patent was issued on March 1, 1994 as Patent #5,290,757 titled "Preemergence Weed Control Using Dipeptides From Corn Gluten Hydrolysate" (Liu et al. 1994). Later work has also shown that there is a pentapeptide, leusine-serine-proline-alanine-glutamine (Leu-Ser-Pro-Ala-Gln), that also has activity (Liu and Christians 1996). An international patent was filed in 1996 in Australia, Canada, Japan, Denmark, France, Germany, Italy, Sweden, Switzerland, and the United Kingdom. This patent includes the five dipeptides covered in the US patent and also includes the pentapeptide. To date, this patent has been allowed a Canadian patent #2,144,321 and Australian patent #679,107 (Christians et al. 1994c).

Site of Action of Dipeptides

While the inhibitory effect of the CGM, the grain hydrolysates, and the peptides could be observed, the exact nature of the inhibition was not well understood. In the next phase of the work, J. Bryan Unruh began studies to observe morphological and anatomical differences in the roots of perennial ryegrass seedlings treated with Ala-Ala, one of the most active dipeptides. Seeds were germinated in Petri dishes containing filter paper wetted with increasing concentrations of Ala-Ala. Root tips from treated seedlings were fixed with 2%:2% paraformaldehyde to glutaraldehyde in 0.1 M Na cacodylate buffer, pH 7.2, at 4°C for 20 h. Root tips were dehydrated in a graded ethanol series, then passed through three changes of acetone, and embedded in resin. Median longitudinal sections, 1 μm and 80 nm thick, were cut on a microtome for light microscopy (LM) and transmission electron microscopy (TEM), respectively. The LM sections were mounted on glass slides, stained with 1% toluidine blue in 1% Na borate for 8-s and then observed microscopically and photographed. TEM sections were placed on 200-mesh copper grids, stained with uranyl acetate and lead citrate and observed and photographed on a scanning transmission electron microscope (Unruh et al. 1997a). The root tips of the treated ryegrass were observed to be devoid of cellular components, specifically discernible nuclei and mitotic structures, with an overall loss of cytoplasmic integrity. The cell walls were observed to have a number of abnormalities that included breakage and uneven thickening.

In a second study, the biological activity of the Ala-Ala was compared to the previously observed effects of commercially available preemergence herbicides (Unruh et al. 1997b). Time-course studies were conducted to observe the mitotic activity of perennial ryegrass seedlings germinated in the presence of Ala-Ala. Ala-Ala exhibited activity on mitosis in root meristems within 4 h of exposure, and by 6 h reduction in the number of mitotic figures was nearly 100%, resulting in only interphase cells. It was concluded that the effect of Ala-Ala on root meristem growth is the result of an inhibition of cell division rather than a disruption of cell division processes. Because only interphase cells were present and no absent or aberrant mitotic stages were noted, we concluded that Ala-Ala is acting on some metabolic process rather than directly on the mitotic apparatus. The observed effect on mitosis caused by Ala-Ala was different from that of the dinitroanilines, carbamates, and dithiopyr, all of which affect the mitotic apparatus and was similar to the chloracetamides and sulfonylurea herbicides, which affect plant metabolic processes.

We also wanted to describe Ala-Ala-induced changes in cellular ultrastructure in an effort to further understand the mode of action of the dipeptide. This was done with light and transmission electron microscopy and again, the results were compared with the reported modes of action of other commonly used herbicides. By 12 h exposure to Ala-Ala, dense droplets, presumably membrane lipids, were visible in vacuoles and intercellular spaces. After 48 h of exposure, epidermal and cortical cells in the treated roots appeared compressed with a disruption in cell polarity. Root lateral branching, similar to effects of preemergence herbicides, was also noted. Root tips showed no gross external abnormalities until after a 96-h exposure to Ala-Ala. The observed effects on cell walls and membranes, as well as the presence of lipid materials in vacuoles and intercellular spaces, implicate Ala-Ala as having membrane-disrupting characteristics. Several classes of herbicides, including the aryl-propanoic acids, cyclohexanediones, thiocarbamates, and chloracetamides, damage cellular membranes by affecting lipid synthesis.

Natural Herbicides

The use of CGM for weed control has been very successful commercially. There are presently 23 companies in the US marketing it as a natural herbicide. It is also now being marketed in Canada and preparation is being made for sales in Ireland and other European countries. The standard program on lawns in cool-season regions includes the application of 10 kg/100 m^2 (20 lbs/1,000 ft^2) in spring before crabgrass germinates, followed by another 10 kg/100 m^2 in late summer to late fall. In addition to the inhibitory effect on germinating weeds, the product provides

2 kg N/100 m^2 (4 lbs N/1,000 ft^2) for the turf. It is also being used for a variety of other horticultural uses including weed control in perennial flower beds, strawberry production, and grapes. It can be used for the production of organic vegetables, although its use is generally limited to crops that are established from transplants. CGM inhibits the establishment of seeded vegetables and is of limited use in seeded gardens and crop production systems (McDade and Christians 2000).

The hydrolysate is currently not being marketed. While CGM is readily available from a variety of producers, no companies are currently producing the hydrolysate in amounts that would be required for national marketing. The hydrolysates are quite expensive to produce and the present lack of alternative markets beyond their use as a natural herbicide has slowed their further development. We have also found that while the hydrolysate is several times more efficacious in laboratory and greenhouse studies, this has not been the case in field studies. The hydrolysate appears to be rapidly degraded by microbial activity in actual field conditions and the herbicidal effect of the hydrolysate is similar to that of the CGM on an equivalent weight basis. This further increases the cost of using the hydrolysate.

In an attempt to improve the economic feasibility of using the hydrolysate, we have been looking for methods of slowing the degradation of the hydrolysate in field conditions to reduce the amount that would be needed. Studies were conducted to observe the effects of soy oil and humic acid on the stability of the hydrolysate in soil. These materials did not significantly improve the efficacy of the material (McDade and Christians 2001). Other methods of protecting the hydrolysate from microbial degradation, such as encapsulation, have been proposed, but have not been tested to date. There is still a lot of interest in the hydrolysates among organic crop producers and it may yet be commercially developed.

The question of whether the peptides can be developed as herbicides is more complex. The market for this type of product would be organic crop producers and those consumers who do not want to use synthetic herbicides. The peptides are extremely expensive to extract from grain and other proteins. They can be synthetically produced but they are then no longer acceptable to the natural product market. Perhaps the peptides can be added to CGM and to hydrolysates to boost their activity, but this work has not been conducted at this time.

References

Christians NE (1991) Preemergence weed control using corn gluten meal. US Patent 5,030,268. (Re-issued with broader claim in April, 1994, Re. 34,594)

Christians NE (1993) The use of corn gluten meal as a natural preemergence weed control in turf. Int Turfgrass Soc Res J 7:284–290

Christians NE, Garbutt JT, Liu D (1994a) Preemergence weed control using plant protein hydrolysate. US Patent No. 5,290,749

Christians NE, Garbutt JT, Liu D (1994b) Preemergence weed control using dipeptides from corn gluten hydrolysate. US Patent No. 5,290,757

Christians NE, Garbutt JT, Liu DL (1994c) Root-inhibiting compounds as growth-regulating compounds and preemergence herbicides. Filed in Australia, Canada, Japan, Denmark, France,

Germany, Italy, Sweden, Switzerland, and the United Kingdom. Inter. Patent App. No. PCT/US94/08513 (Canadian Patent # 2,144,321 Preemergence Weed Control Using Natural Herbicides. Granted July 1999)
Liu DL, Christians NE (1994) Isolation and identification of root-inhibiting compounds from corn gluten hydrolysate. J Plant Growth Regul 13:227–230
Liu DL, Christians NE (1996) Bioactivity of a pentapeptide isolated from corn gluten hydrolysate on *Lolium perenne* L. J Plant Growth Regul 15:13–17
Liu DL, Christians NE (1997a) Inhibitory activity of corn gluten hydrolysate on monocotyledonous and dicotyledonous species. Hortscience 32:243–245
Liu DL, Christians NE (1997b) The use of hydrolyzed Corn Gluten Meal as a natural preemergence weed control in Turf. Int Turfgrass Soc Res J 8:1043–1050
Liu DL, Christians NE, Garbutt JT (1994) Herbicidal activity of hydrolyzed corn gluten meal on three grass species under controlled environments. J Plant Growth Regul 13:221–226
McDade MC, Christians NE (2000) Corn gluten meal-a natural preemergence herbicide, Effect on vegetable seedling survival and weed cover. Am J Alternat Agr 15(4):189–191
McDade MC, Christians NE (2001) Corn gluten hydrolysate for crabgrass (*Digitaria* spp) control turf. Int Turfgrass Soc Res J 9:3–7
Unruh JB, Christians NE, Horner HT (1997a) Herbicidal effects of the dipeptide Alaninyl-Alanine on perennial ryegrass (*Lolium perenne* L.) seedlings. Crop Sci 37:208–221
Unruh JB, Christians NE, Horner HT (1997b) Mitotic and ultrastructural changes in root meristems of grass seedlings treated with Alaninyl-Alanine. Crop Sci 37:1870–1874

Chapter 9
Physiological Importance and Mechanisms of Protein Hydrolysate Absorption

Brian M. Zhanghi and James C. Matthews

Abstract Understanding opportunities to maximize the efficient digestion and assimilation by production animals of plant- and animal-derived protein products is critical for farmers, nutritionists, and feed manufacturers to sustain and expand the affordable production of high quality animal products for human consumption. The challenge to nutritionists is to match gastrointestinal tract load to existing or inducible digestive and absorptive capacities. The challenge to feed manufacturers is to develop products that are efficient substrates for digestion, absorption, and/or both events. Ultimately, the efficient absorption of digesta proteins depends on the mediated passage (transport) of protein hydrosylate products as dipeptides and unbound amino acids across the lumen- and blood-facing membranes of intestinal absorptive cells. Data testing the relative efficiency of supplying protein as hydrolysates or specific dipeptides versus as free amino acids, and the response of animals in several physiological states to feeding of protein hydrolysates, are presented and reviewed in this chapter. Next, data describing the transport mechanisms responsible for absorbing protein hydrolysate digestion products, and the known and putative regulation of these mechanisms by their substrates (small peptides) and hormones are presented and reviewed. Several conclusions are drawn regarding the efficient use of protein hydrolysate-based diets for particular physiological states, the economically-practical application of which likely will depend on technological advances in the manufacture of protein hydrolysate products.

Keywords Milk by-products • Ovalbumin • Protein absorption • Peptides • PepT1 • Substrate regulation • Hormonal regulation

---

B.M. Zhanghi and J.C. Matthews (✉)
Department of Animal Science, University of Kentucky, Lexington, KY, USA
e-mail: jmatthew@uky.edu

V.K. Pasupuleti and A.L. Demain (eds.), *Protein Hydrolysates in Biotechnology*,
DOI 10.1007/978-1-4020-6674-0_9,

Introduction

Nutritionists constantly strive to formulate diets and develop feeding regimens that maximize nutrient assimilation and minimize nutrient excretion. Within a given species, the relative efficiency of nutrient assimilation changes in normal (e.g., in utero through adult growth, gestation and lactation, adult maintenance and aging) and abnormal (e.g., starvation, disease, intentionally-induced long term metabolic stress) physiological states. Different assimilation efficiencies also exist for nutrient classes among and within species that utilize different digestive strategies (ruminants versus nonruminants, humans versus dogs, goats versus cattle). However, for each of these physiological states and species, the efficient assimilation of nutrients that enter the gastrointestinal tract as polymers of constituent nutrients (e.g., starch, protein, triglycerides), depends on the ability of the animal to express functional digestive capacities that match or exceed corresponding absorptive capacities. In contrast, the assimilation of nutrients that appear in digesta as absorbable units (e.g., certain complexed and/or free minerals and vitamins, FAAs and small peptides) depends primarily on absorptive capacity.

Protein assimilation is a multistage process that includes gastric, intestinal lumen, glycocalyx/apical membrane, and intracellular hydrolytic events, and mediated (primarily) absorption of peptides and FAAs (FAA) across both apical and basolateral membranes of absorptive epithelial cells (enterocytes). A working model for assimilation of luminal protein hydrolysates by enterocytes that emphasizes enterocytic events is illustrated in Fig. 1, whereas more comprehensive models and discussions of protein assimilation are available (e.g., Matthews 1991; Ganapathy et al. 1994; Gardner 1994; Krehbiel and Matthews 2003). Peptides are absorbed from gastrointestinal digesta by the activity of PepT1 (Fei et al. 1994). PepT1 mediates the H^+-coupled cotransport of peptides, and displays a wide range of affinity (over 100-fold) among peptides and certain classes of natural and designed pharmaceutical compounds (Meridith and Boyd 2000). In the gastrointestinal tract tissues of nonruminants and ruminants, the primary site of PepT1 expression is the apical membrane of enterocytes. Following absorption, about 70–80% of all absorbed peptides are hydrolyzed by intracellular peptidases, whereas non-hydrolyzed peptides cross the basolateral membrane by an unidentified peptide transporter(s) that mediates the non-H^+-coupled passage of peptides. In contrast to peptides, the passage of FAA across apical and basolateral membranes is mediated by a greater variety of transport proteins, each of which possesses a greater selectivity of substrates than does PepT1 (Palacin et al. 1998; Krehbiel and Matthews 2003).

For efficient dietary protein assimilation, digestive and absorptive capacities of gastrointestinal tract tissues must be matched. Traditionally, inadequate protein assimilation from a given diet to meet the AA requirements for a given physiological state have been compensated for by formulating diets that contain supplemental FAA. However, the relatively recent understanding that the majority of AAs feedstuffs are absorbed as di- and tripeptides (peptides), and not as FAAs, has resulted in research to assess the benefit of supplying supplemental AAs in the form of peptides.

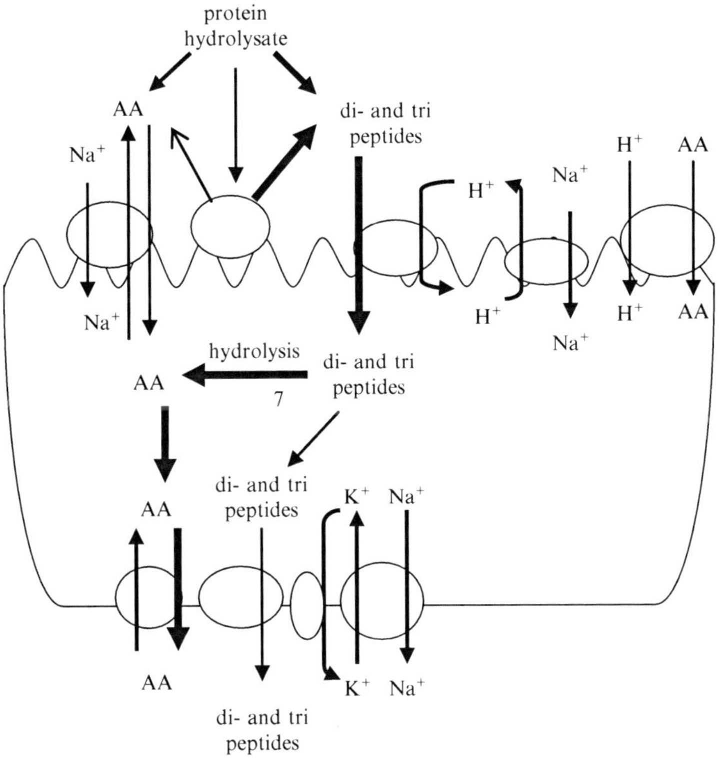

Fig. 1 Model for assimilation of protein from luminal hydrolysates (Adapted from Ganapathy et al. 1994; Krehbiel and Matthews 2003)

In this chapter we examine literature about the (1) capacity of gastrointestinal tract epithelial tissues to assimilate dietary AAs as free versus peptide-bound AAs, (2) influence of peptide hydrolysates and bioactive dietary peptides on animal performance, (3) biochemical characterization of PepT1 activity, and (4) potential to regulate PepT1 functional capacity and related proteins that support PepT1 function. We conclude with thoughts about the future use of protein hydrolysates and/or peptides as dietary constituents.

Results and Discussion

Comparison of Free Versus Peptide-Bound AA Absorption Capacity

The relative amount of α-amino N absorbed from the digesta is related to the extent and rate of hydrolysis of protein to small peptides and FAAs during digestion. Given this understanding, and that small peptides constitute the major form of AA absorption from digesta, it would be expected that feeding protein as hydrolysates rather than intact or FAA would increase total AA absorption, assuming that peptide absorption capacity increased proportionally with digesta levels. The influence of feeding hydrolysates compared to FAA mixtures containing a comparable AA profile have been evaluated primarily in humans subjects, whereas comparison of protein hydrolysates (peptides) to an intact protein source has been evaluated in non-human animals, such as pigs, rats and dogs. A summary of these studies is presented in Tables 1 and 2. In general, delivery of AAs into the jejunum as peptides results in more moles of AA absorbed as peptides rather than as FAA, when total AA concentrations range from 30 to 64 mM (Table 1).

The threshold for enhanced luminal N-disappearance of peptides versus FAA is achieved when the relative abundance of peptides as di- and tripeptides is at least 46% of the α-amino N, and the corresponding concentration of di- and tripeptides ranges from 4.6 to 6.9 mM. Additionally, this peptide concentration range appears to increase with greater protein loads, from 70 to 100 mM. When the protein load is at these higher levels, enhanced N-disappearance with peptides compared to FAA occurs when the concentration of di- and tripeptide ranges from 10.7 to 16 mM. Paralleling this observation is that the concentration ranges of peptides used were at or above the K_m of most peptides recognized by PepT1. In contrast, the concentration of equimolar amounts of AAs as FAA exceeded the typical K_m (μM) of known (Krehbiel and Matthews 2003) intestinal AA transporters.

In general, studies that have evaluated the disappearance of protein N when delivered as protein hydrolysates versus intact protein or FAA also revealed that protein hydrolysates seemed to be the more effective way to supply AAs, if the protein load was large enough (Table 2). This conclusion is exemplified by a study with dogs (Zhao et al. 1997), which showed that intestinal N-disappearance was 13.5% greater when AAs were supplied as protein hydrolysates rather than as intact protein. Unfortunately, no description of the peptide fraction was described, so it is difficult to assess the lack of an effect on N-disappearance at the lower protein levels. However, the researchers did state that the experimental hydrolysate was free of intact protein and FAA.

Although the functional capacity of PepT1 and FAA transporters has not been directly compared, the data summarized in Tables 1 and 2 indicate that the enhanced luminal N-disappearance of AAs when supplied in the form of peptides resulted from a singularly greater PepT1 activity versus that of the complement of apical membrane FAA transporter activities. However, it is important to consider the possibility that the "sole" PepT1 activity may have in fact been contributed to by FAA transporters, after hydrolysis of peptides by apical membrane peptidases. Consistent with this concept, relative changes in the disappearance of specific AAs when AAs were supplied as protein hydrolysates versus FAA have been observed, if total AA loads are increased (Table 3). In general, these data suggest that His, Tyr, Phe, Lys, Thr, Glu, Asp, Ser, and Ala are absorbed in greater amounts when present in the small intestinal lumen as peptides than as FAA. In contrast, the AAs Met, Arg, and Pro generally appear to be absorbed in greater amounts when supplied as FAA.

These in vivo (N-disappearance trials of Tables 1 and 2) data appear to complement our understanding of known intestinal AA transport activities (Matthews 2000a)

Table 1 Comparison of nitrogen (N) disappearance from human intestinal lumen after intestinal infusion of protein as hydrolysate versus FAAs

Protein type[a]	TAA infused[b], mM	Di- and tripeptides,%[c]	Di- and tripeptides, mM[d]	N-disappear[e]	Citation
Ovalb	30	24	2.4–3.6	--	Grimble et al. 1987a
Lactalb	40	32	4.3–6.4	0	Hegarty et al. 1982a
Ovalb	30	46	4.6–6.9	+	Grimble et al. 1987a
Ovalb	30	75	7.3–11.0	+	Grimble et al. 1987a
Casein	40	66	8.8–13.2	+	Silk et al. 1973a
Casein	40	66	8.8–13.2	+	Silk et al. 1980
Casein	64	50	10.7–16.0	++	Silk et al. 1973a
Casein	64	50	10.7–16.0	++	Silk et al. 1980
Lactalb	100	0	0	---	Grimble et al. 1987b
Ovalb	100	24	8–12	-	Grimble et al. 1987a
Lactalb	70	32	7.5–11.2	0	Hegarty et al. 1982a
Lactalb	100	32	10.7–16.0	+	Grimble et al. 1987b
Lactalb	100	32	10.7–16.0	++	Hegarty et al. 1982a
Ovalb	100	46	15.3–23.0	++	Grimble et al. 1987a
Lactalb	70	80	18.7–28.0	+++	Silk et al. 1980
Ovalb	100	75	25.0–37.5	+	Grimble et al. 1987a
Fish	70	72	16.8–25.2	0	Silk et al. 1980

[a]Ovalb: ovalbumin; Lactalb: lactalbumin.

[b]Total AA (TAA) after infusion with protein hydrolysate or with equimolar amounts of FAAs.

[c]Relative portion of alpha-amino N as di- and tripeptides present in the infused protein hydrolysate.

[d]Estimated millimolar range of di- and tripeptides present in the infused protein hydrolysate. Average values determined by calculating the millimolar amount of AAs in the di- and tripeptide form, then dividing by 2 (di-peptide) or 3 (tripeptide). For example, 100 mM TAA × 0.32(% as di-tripeptides) = 32 mM as AA in di- and tripeptide form, 32 mM/2 = 16.0 mM dipeptides or 32 mM/3 = 10.7 mM tripeptides.

[e]Relative change in N disappearance from intestinal lumen; + = 10–25%; ++ = 25–50%; +++ = >50%; 0 = no difference (–10% to 10%); - = –10% to –25%; -- = –25% to –50%; --- = < –50%. TAA: total AAs.

Table 2 Comparison of nitrogen (N) disappearance from animal intestinal lumen after intestinal perfusion of protein as a hydrolysate versus intact protein or FAAs[a]

Species	Protein type	Amount infused	Protein form[b]	Sample time, m	N disap.[c, d]	Citation
Pig	Casein	16 g/L	H versus FAA	12 h	-, a	Buraczewska 1981
50–75 kg		16 g/L			-, b	
		16 g/L			0, c	
14 days old	Milk	15% CP	H versus I		0	Moughan et al. 1990
Canine	Soy	50 g/L	H versus I		0	Defilippi and Cumsille 2001
	Casein	50 g/L	H versus I		0	
	Soy	100 g/L	H versus I	60–120	0	Zhao et al. 1997
		200 g/L			+	
		100 g/L	H versus I		0	
		200 g/L			+	
Rat	Whey	10 mg	H versus I	60	0	Ziemlanski et al. 1978
			H versus I		++	

[a]Total AA levels are reported as N disappearance from perfusate after perfusion with equimolar amounts of protein hydrolysate (H), free AA mixture (FAA), or intact protein (I).
[b]Peptide fractionation of hydrolysate not described.
[c]Relative change in N disappearance: + = 10–25%; ++ = 25–50%; +++ = >50%; 0 = no change (–10% to 10%); - = –10% to –25%; -- = –25% to –50%; --- = < –50%.
[d]Site of infusion: a: Du = duodenum, b: Je = jejunum, c: I = ileum.

and proteins (Krehbiel and Matthews 2003) expressed in the apical membrane of small intestinal absorptive cells. For example, the aromatic AAs Tyr and Phe typically demonstrate the largest absorption capacity when supplied as hydrolysates at nearly all levels of protein load (Silk et al. 1975; Hegarty et al. 1982a; Grimble et al. 1987a; b). This phenomenon probably results from the lack of expression of a specific aromatic AA transporter on the apical membrane, the fairly high recognition of aromatic-containing peptides by PepT1, and the expression of a specific aromatic AA transporter (TAT1) on the basolateral membranes of enterocytes. That is, once absorbed by PepT1 as peptides into the cell, and cleaved by intracellular peptidases, the released Tyr and Phe are readily transported out of the cell by the low-affinity (mM) but high-capacity TAT1.

Similarly, if the relative affinities and capacities of PepT1 and system X_{AG}^- anionic transporters are considered together, then a plausible hypothesis to explain the observations that more Glu and Asp are absorbed when supplied as hydrolysates of casein as compared to FAA is readily developed. That is, although two system X_{AG}^- transporters (EAAC1, GLT-1) are expressed in at least the apical membrane of enterocytes, these proteins are high-affinity (low μM) transporters that appear to be expressed in relatively low abundance. Although PepT1 also appears to possess a relatively high-affinity for anionic AA-containing peptides, it also possesses a high capacity for transport, unlike EAAC1 and GLT-1. Accordingly, in the presence of relatively low total Glu and Asp, the capacity of EAAC1 and GLT-1 are not exceeded and relative differences between the amount of substrate transported by PepT1 and the EAAC1 and/or GLT-1 are not discernible. However, when anionic AA concentrations increase, the high capacity PepT1 has the potential to transport more moles of Glu and Asp than do the system X_{AG}^- transporters. This hypothesis is strengthened by the knowledge that the two studies (Silk et al. 1973b; Silk et al. 1975) that observed the enhanced uptake of Glu and Asp when supplied as hydrolysates were conducted using casein, which contains relatively high levels of Glu.

Appearance of AAs and Peptides in Blood

Certain digesta-derived AAs are extensively used to support the dynamic metabolic requirements of the intestinal epithelium. A great deal of work has been performed to derive this general understanding, including studies that have attempted to measure the net balance of AA metabolism, by quantifying the arterial-venous (A-V) flux of AAs across the intestinal epithelium. Besides being transported across the basolateral membrane into blood, absorbed AAs are oxidized for energy (Glu, Asp, and Gln; Windmeuller and Spaeth 1980; Wu 1998), converted into nucleic acids (Asp and Gln; Burrin and Reeds 1997) or glutathione (Glu; Reeds, et al. 1997), and incorporated into de novo protein synthesis. Consequently, the appearance of AAs in portal blood is not reflective of the digestive (nor dietary disappearance) AA profile (Stoll et al. 1998). A-V flux studies in pigs (Rerat and Nunes 1989; Rerat et al. 1992) also have shown that the dietary form of protein can dramatically affect postprandial portal appearance of total (Table 3) and individual (Table 4) AA acids.

Table 3 Relative change in disappearance of individual FAAs when protein source is administered as a protein hydrolysate versus FAAs[a]

Species	Human	Human	Human	Human	Human	Human	Human	Human
Hydroy-sate	Cas[b]	Cas[c]	Lactalb[d]	Ovalb[e]	Ovalb[e]	Lactalb[f]	Lactalb[f]	Lactalb[f]
Protein amount	40 mM	64 mM	100 mM	30 mM	100 mM	40 mM	70 mM	100 mM
Sample time, min	Not given	60	60	30	30	30–60	30–60	30–60
His	1.25x[x]	2x	2.4x[x]	1	0.7x	1.1x	0.9x	2.1x[x]
Lys	1.3x[x]	1.2x	1.1x	1	0.8x	1	1	1.2x
Phe	1.4x[z]	1.7x[z]	2x[y]	1.2x	1.4x[x]	1	1	1.3x[x]
Leu	0.85x	1	1.2x[x]	1	1	0.9x[x]	0.8x	1
Ile	0.9x	1	1	0.7x	0.8x	0.9x[x]	0.7x[x]	1.1x
Met	0.75x[*]	1	1.1x	1	1.1x	0.8x[x]	0.9x[x]	0.9x
Val	1	1.1x	1.6x[x]	1.1x	1.1x	0.8x	0.8x[x]	1.2x
Thr	1.1x	1.6x[y]	2x[x]	1.3x	1.5x	1	0.9x	1.8x[x]
Arg	1	1	1	1	1	0.6x[x]	1	0.8x
Cys			1					
Tyr	1.1x	1.7x	1.5x	1.4x	2.2x[x]	1	1	1.4x[x]
Asx			2.8x[y]	1	1			
Asp	2.1x[y]	2x[x]						
Asn						2.2x[x]	2.7x[x]	2.5x[x]
Pro	1	1.2x	0.9x	1.2x	1	0.8x[x]	0.6x[x]	0.7x
Ser	1.25x	1.5x[x]	2x[y]	1.3x	1.5x[x]	0.9x	0.8x	1.5x[x]
Glx			2x[y]	1.8x	1.3x			
Glu	1.3x[y]	1.4x						
Gln						1.2x	1.3x[x]	2.2x[x]
Gly	1.1x[*]	1.1x	1.3x[x]	1.1x	0.9x[x]	1	0.9x	1.2x

Ala	$1.3x^{x}$	$1.7x^{x}$	$1.5x^{y}$	1.4x	1.3x	0.9x	0.9x	1.2x
Trp				0.9x	2.9x			
Citation	Silk et al. 1973b	Silk et al. 1975	Grimble et al. 1987b	Grimble et al. 1987a	Grimble et al. 1987a	Hegarty et al. 1982a	Hegarty et al. 1982a	Hegarty et al. 1982a

[a]Values are the amount of a given AA in protein hydrolysate perfusate, relative to that versus a mixture of FAAs (FAA), that disappears from the jejunal digesta after perfusion: * $P<0.10$ $^{x}P<0.05$ $^{y}P<0.01$ $^{z}P<0.001$.
[b]Casein: 66% di- and tripeptides, 34% FAA.
[c]Casein: 50% di- and tripeptides, 50% FAA.
[d]Lactalbumin: 32% di- and tripeptides, 63% 4–5 AA, 5% FAA.
[e]Ovalbumin: 75% di- and tripeptides, 16% 4–5 AA, 9% FAA.
[f]Lactalbumin: 30–35% as di-, tri-, and tetrapeptides, 20% FAA, 20–30% > tetrapeptide and < 1,500 MW, 10–20% > 1,500 MW.

Table 4 Relative change in total AA concentration of plasma when protein is administered as a protein hydrolysate versus as a free AA or intact protein[a]

Species	Protein type	Protein form	Protein amount	Blood sample	Sampling time, min	Total AA levels[d]	Citation
Pig	Milk	H[e] versus FAA	55 g[b]	Portal	60	+++	Rerat and Nunes 1989
57 kg			110 g[b]		300	++	
				A-V flux	60	+++	
					300	+++	
57 kg	Milk	H[e] versus FAA	110 g[b]	Portal A-V flux	60	++	Rerat et al. 1992
1 day old	Soy	H[f] versus I	25% CP	Portal	15	0	Zijlstra et al. 1996
					30	0	
					60	0	
					120	0	
				Cardiac	15	-	
					30	0	
					60	0	
					120	++	
Rat 240–260 g	Egg White	H[g] versus FAA	18% CP[b]	Portal[n]	10	++	Hara et al. 1984
60–80 g	Milk	H[e] versus FAA	1 g[b]	Cardiac	60	0	Monchi and Rerat 1993
	Whey	H versus I		Portal	7	+	Nakano et al. 1994a
					30	-	
Weanling	Whey	H[h] versus I	12% CP[c]	Arterial[o]	7 days feeding	0	Baro et al. 1995
	Casein	H[i] versus I				0	
Guinea pig 250–550 g	Casein	H[j] versus FAA	120 mM[b]	Portal	30	-	Sleisenger et al. 1977
					60	0	

Human 20–30 years	Casein	H[k] versus FAA	10 g[c]	Venous[p]	15	0	Marrs et al. 1975
					30	0	
					45	0	
Adult	Fish	H[l] versus FAA	50 g[c]	Venous[p]	30	+++	Silk et al. 1979
					60	++	
					120	0	
					180	-	
19–32 years	Lactalb	H[m] versus A	50 g[c]	Venous[p]	30	++	Hegarty et al. 1982b
					60	++	
					120	++	
					180	0	

[a]Relative difference in AAs after perfusion with protein hydrolysate versus perfusion with FAAs or intact protein (I); + = 10–25%; ++ = 25–50%; +++ = >50%; 0 = no difference (–10% to 10%); - = –10% to –25%; -- = –25% to –50%.
[b]Duodenal infusate.
[c]Orally ingested.
[d]Note.
[e]Milk: 60% < pentapeptide, 20% pentapeptide to 1,500 MW, and 17% 1,500 < MW < 5,000.
[f]Soy hydrolysate: 33% 100–1,500 MW, 5% FAA, 62% > 1,500 MW.
[g]Egg white with average MW of 350, >70% di- and tripeptides, <10% as FAA.
[h]Whey: 44.5% 200–800 MW, 3% FAA, 44.3% 800–8,000 MW.
[i]Casein: 72.5% 200–800 MW, 14.6% FAA, 13.3% 800–2,500 MW.
[j]Casein: >50% di- and tripeptides.
[k]Casein:~66% di- and tripeptides, 33% FAA.
[l]Fish: 72% di- and tetrapeptides, 20% FAA, 8% > tetrapeptides.
[m]Lactalbumin: 30–35% as di-, tri-, and tetrapeptides, 20% FAA, 20–30% > tetra-peptide and <1,500 MW, 10–20% > 1,500 MW.
[n]Portal serum was analyzed for alpha amino nitrogen.
[o]Arterial blood collected after decapitation.
[p]Peripheral venous blood.

Specifically, investigations utilizing catheterized pigs have provided tremendous advances in our understanding of the relative nutritional contribution of peptide hydrolysates, and have been reviewed in detail by others (Rerat 1993, 1995). Therefore, the studies included in Table 4 will only be briefly discussed to underscore the findings as they relate to AA and peptide transport. Jejunal infusions of either 55 (6.2% of diet) or 110 g (12.5% of diet) of protein has demonstrated that absorption of AAs from the peptide hydrolysate, as determined by the increased appearance of total AAs in portal blood, occurred earlier and at a greater rate than from protein supplied as FAA (Rerat et al. 1988). After 60 min, the amount of total AAs appearing in portal blood was approximately 75% and 60% greater when infused as peptide compared to FAA at 55 and 110 g, respectively. At 5 h (300 min), these relative differences changed to 28.7% and 100%, remaining in favor of peptide infused at 55 and 100 g, respectively. Additionally, comparison of infusion of 55 or 110 g of FAA indicated that approximately the same total amount of AAs appeared in the portal vein after 5 h, indicating a saturation of AA absorption capacity. In contrast, infusion of peptide hydrolysate demonstrated a 45.2% increase in AA appearance (peptide absorption capacity) when the amount infused was increased from 55 to 110 g.

A follow-up study was performed to investigate the inclusion of 440 g carbohydrates (maltose-dextrin) with 110 g of either hydrolysate or FAA on appearance of AAs in portal blood (Rerat et al. 1992). Similarly, the hydrolysate infusion resulted in a 34.9% greater appearance of total portal AAs versus FAA infusion. However, infusing the hydrolysate with the maltose-dextrin resulted in a decrease (37.9%) in total cumulative portal AA appearance compared to previous experimental observations (Rerat et al. 1988). This suggests a nutrient interaction between carbohydrates and peptides, possibly resulting from a change in the viscosity of the unstirred layer and/or increased digesta passage, causing reduced peptide uptake. Similar effects were observed with an increase in dietary cellulose, in which an increase from 48 (6% cellulose) to 128 g (16% cellulose) demonstrated a 25% decrease in N-digestibility and a 26.2% decrease in total α-amino N appearing in portal blood over 12 h following intake (Giusi-Perier et al. 1989).

In contrast to studying changes in intestinal A-V flux, other research has investigated changes in only portal (Hara et al. 1984; Sleisenger et al. 1977) or arterial (Monchi and Rerat 1993) blood after duodenal infusion in rats and guinea pigs, and conflicting results have been observed by examining these blood pools separately. Generally, rats appear to demonstrate a greater uptake capacity for peptides compared to guinea pigs, as indicated by a greatly enhanced appearance of portal AAs from a protein hydrolysate versus FAA. In both these studies, di- and tripeptides constituted greater than 50% of the hydrolysate infused duodenally, therefore, why the inconsistency? Possibly, the form of protein may be less of a factor compared to a difference in the amount, type of protein, peptide profile, or just species difference.

The profile and total amount of AAs that appear in the portal circulation following a meal are available for further metabolism by hepatic tissue. With the exception of the branched chain AAs (Leu, Ile, Val) that are not readily metabolized by hepatocytes because of a lack in expression of BCAA aminotransferase (Torres et al. 1998), the liver is known to play a critical role in regulating the levels of FAAs in the plasma of the peripheral circulation (Kilberg and Haussinger 1992). Consequently, AAs in the peripheral circulation are available for uptake and metabolism by peripheral tissue for maintenance and growth. Several human studies have been performed to compare the feeding of protein hydrolysates and FAA to characterize the temporal response of AA levels in peripheral venous blood. The findings of these studies are generally summarized in Table 3 and indicate that meals with protein as hydrolysate will enhance the total appearance of AAs in venous blood. However, this appears to be directly influenced by the dietary load, as 10 g of hydrolysate did not demonstrate a difference compared to FAA (Marrs et al. 1975), whereas 50 g enhanced total AA appearance from 30 to 120 min following ingestion (Silk et al. 1979; Hegarty et al. 1982b).

In addition to FAA appearing in the portal and peripheral blood following intestinal perfusion of hydrolysates, di- and tripeptides absorbed from the small intestinal lumen also appear intact as peptide-bound AAs (Gardner 1984). Early research investigating the transepithelial passage of intact peptides has demonstrated that 10–30% of the peptide-bound AAs absorbed into small intestinal epithelial cells will appear as intact peptides in the serosal secretion fluid or mesenteric blood of rats or guinea pigs, resulting from intestinal perfusion of casein or soybean protein hydrolysates, or specific peptides (Gly-Phe, Leu-Gly; Gardner 1982; Gardner et al. 1983). Current research has provided further evidence of intact peptides enter the peripheral circulation of dogs following a single oral bolus of glycylsarcosine (GlySar), with the observation that 11.5% of absorbed GlySar appeared in serum in the peptide-bound form (Zanghi et al. 2004). Additionally, β-Ala-His (carnosine) absorption was observed in humans following a single oral bolus; however detection of carnosine was largely restricted to the presence of intact peptide in urine, where 14% of the ingested load was recovered (Gardner 1982). Further evaluation ultimately indicated the presence of carnosine in blood, but hydrolysis was rapid, with a blood appearance half-life of approximately 2 min at 37°C. Although there is evidence for appearance of intact peptides in the blood following absorption from the intestine, the ultimate nutritional significance of peptides derived from intestinal absorption for total N-balance is still unclear. In contrast, digestion-derived bioactive di- and tripeptides that possess therapeutic properties, would be of considerable significance and is discussed later in the chapter.

Although the contribution of peptides appearing in the mesenteric blood following absorption appears to be a much smaller proportion versus FAA, this does not exclude the existence of large portions of α-amino N present in the circulation in the peptide-bound form as di- and tripeptides. This topic has been reviewed in detail previously (Krehbiel and Matthews 2003). In brief summary, several studies have demonstrated that 63–92% of α-amino N in the portal drained visera of sheep and cattle (McCormick and Webb 1982; Seal and Parker 1991; Koeln et al. 1993; Han et al. 2001), and 52% in rats (Seal and Parker 1991), was in the peptide bound form. Recent research has also indicated that peripheral tissues (mammary) in dairy cattle utilize peptide-bound AAs specifically, in addition to FAA (Tagari et al. 2004). Therefore, in addition to FAA, it will be important that future studies evaluate all pertinent blood pools for shifts in peptide-bound AAs in response to hydrolysate feeding.

Influence of Hydrolysate or FAA on Individual AA Portal and Peripheral Blood Appearance

As reviewed above, hydrolysates enhance the total AA load appearing in portal and peripheral blood. However, it is important to note that the appearance of specific AAs are not equally enhanced, especially for AAs that may be conditionally or chronically limiting for a given physiological state. The relative increase or decrease in the level of individual AAs in blood following feeding of a protein as hydrolysate or FAA is detailed in Table 5. In particular in pigs, portal levels of His, Lys, Phe, Leu, Thr, Arg, and Tyr demonstrate marked increases at 60 min post-infusion when delivered as hydrolysates (Rerat and Nunes 1989; Rerat et al. 1992), whereas Met was regularly observed to have a reduced portal appearance compared to FAA. In comparison to the AA disappearance data (Table 2), the increased intestinal disappearance of His, Lys, Phe, Thr, and Tyr in humans matches the relative elevation in appearance of these same AA in portal blood of pigs. This suggests that apical peptide transport, intracellular hydrolysis, and basolateral AA transport capacities of pigs and humans are similar. Therefore, pigs may be a good model to study human intestinal AA absorption events. Furthermore, over a period of 30–120 min after infusion, Lys, His, Tyr, Ile, and Leu continue to be observed at elevated levels in human peripheral blood, with the addition of Met and Val, after intestinal infusion of hydrolysate compared to FAA. Overall, there appears to be a positive relationship between enhanced absorption of specific AAs and their elevated appearance in peripheral blood, and therefore, an increased availability for uptake by non-visceral tissues.

It is well known by nutritionists that surplus AAs can be metabolized to their constituent α-keto acids to act as an energy substrate, and amines which are converted to urea and excreted (Heger 2003). However, flux studies across peripheral tissues in pigs indicate that intestinal infusion of hydrolysates result in a greater cumulative peripheral tissue uptake of Lys, Leu, Phe, Ile, Thr, Ser, and Pro and also reduced blood urea nitrogen versus infusion of FAA (as reviewed by Rerat 1995). Interestingly, release of Gln and Ala from peripheral tissues was lower (16% and 60%, respectively) with the hydrolysate infusion as compared to the FAA perfusion. This suggests that the elevated uptake of AAs, especially the branched chain AA, does not result in immediate catabolism, but instead is likely incorporated into protein. Because the profiles of the increased FAA in peripheral circulation of humans and cumulative uptake in pigs were similar, the use of pigs to evaluate

Table 5 Relative change in individual AA levels in plasma when protein source is administered as a hydrolysate versus free AA[a]

Species	Pig	Pig	Pig	Rat	Human	Human	Human	Human
Protein type	Milk[b]	Milk[b]	Milk[b]	Milk[b]	Lactalb[c]	Lactalb[c]	Lactalb[c]	Lactalb[c]
Protein	55 g[d]	110 g[d]	110 g[d]	0.16 g[d]	50 g[e]	50 g[e]	50 g[e]	50 g[e]
Amount	Portal	Portal	Portal	Cardiac	Venous	Venous	Venous	Venous
Blood site	AV flux	AV flux	AV flux					
Sample time, min	60	60	60	60	30	60	120	180
His	4x[y]	3.5x[y]	2.1x[y]	0.8x	1.3x	1.3x	1.2x	0.5x
Lys	2x[y]	2x[y]	1.5x[y]	1	1.2x	1.4x	1.3x	1.4x
Phe	1.9x[x]	2.6x[y]	1.8x[x]	1.1x	1	1.1x	0.9x	1.1x
Leu	1.3x	1.4x	1.2x	1.6x[y]	1.4x	1.5x	1.2x	1
Ile	1x	1.1x	1.3x	1.1x	1.4x	1.5x	1.3x	1.1x
Met	0.5x[x]	0.6x[x]	0.8x	1.3x	1.4x	1.3x	1.1x	1.1x
Val	1.2x	1.1x	0.9x	1.1x	1.4x	1.2x	1.3x	1
Thr	3.5x[y]	2x[y]	1.6x[x]	0.6x	1.1x	1.1x	1.3x	1
Arg	1.8x[x]	2.2x[x]	2.0x[y]	1.6x[y]	0.9x	1.3x	1	0.9x
Cys	1	1	0.5x	2.1x[f,y]				0.8 x
Tyr	3.6x[y]	7x[y]	2.9x[y]		1.3x	1.7x	1.5x	
Asx	1	1		3x[y]				
Asp			0.7x					1.4x
Asn				1	2.0x[x]	1.7x	1.2x	1
Pro	1.2x	1.4x	2.3x[y]	4x[y]	1.2x	1	1.1x	
Ser	2.5x[y]	2.3x[y]	1.3x	0.6x[x]				
Glx	2.8x[y]	2x[x]	1	0.7x				
Glu				1				1.1x
Gln				1.1x	1.9x[x]	1.7x[x]	5.7x	0.5x
Gly	2x[x]	1.8x	1	0.7x	0.9x	1.1x	2.0x	1
Ala	2x[y]	2.2x[y]	1.3x	0.9x	1.2x	1.5x	1.3x	
Trp								

(continued)

Table 5 (continued)

Species	Pig	Pig	Pig	Rat	Human	Human	Human	Human
Citation	Rerat and Nunes 1989	Rerat and Nunes 1989	Rerat et al. 1992	Monchi and Rerat 1993	Hegarty et al. 1982b	Hegarty et al. 1982b	Hegarty et al. 1982b	Hegarty et al. 1982b

[a]Values are the amount of a given AA in protein hydrolysate perfusate, relative to that versus a mixture of FAAs, that disappears from the jejunal digesta after perfusion: $^{x}P<0.05$ $^{y}P<0.01$.

[b]Milk: 80% < 1,500 MW (60% < 5AA residues) and 17% (1,500 < MW < 5,000).

[c]Lactalbumin: 30–35% as di-, tri-, and tetrapeptides, 20% FAA, 20–30% > tetrapeptide and < 1,500 MW, 10–20% > 1,500 MW.

[d]Administration as duodenum infusion.

[e]Administration by oral ingestion.

[f]Represents changes in measured cystine levels.

whether feeding protein hydrolysates (versus FAA) to humans improves net protein utilization by peripheral tissues seems justified.

Influence of Peptide Hydrolysates on Animal Performance

Predigested protein sources have demonstrated increased intestinal disappearance and appearance of nitrogen in portal and peripheral blood in healthy rats, pigs, dogs, and humans. Therefore, it is important to investigate the corresponding physiological impact of feeding a protein hydrolysate to enhance "whole-body performance". For both production and companion animals, whole-body performance is a subjective evaluation and varies widely to accommodate a particular condition or physiological state. For the healthy production animal, maximizing daily feed intake and daily weight gain for "optimal" accretion of tissue mass with the minimum cost is the ultimate goal, whereas feeding the healthy adult companion animal in the maintenance state targets no weight loss or gain, and strives for maximum health to promote longevity of lifespan. Contrary to the wealth of research that has been performed to evaluate whole-body animal performance in response to feeding various types of intact protein sources, studies investigating the feeding of protein hydrolysates on changes in animal performance have not been widely performed in the healthy animal.

In addition to the normal functioning intestinal tract, oral delivery of nutrients is a significant concern when digestive function is compromised as a result of weaning, infection, injury, or surgery. Under these circumstances, a reduction in intestinal digestive capacity leads to a malnourished state. The use of protein hydrolysates in dietary formulations have been implicated in the improvement of nitrogen status during malnourished conditions compared to feeding protein in the intact or FAA form.

Likewise, protein hydrolysates have pharmacological effects from bioactive peptides derived from proteolysis of various proteins Yamamoto et al. 1985). This has also led to investigations on the efficacy of particular di- and tripeptides generated from enzymatic digestion to facilitate a "nutraceutical-like" therapeutic action towards conditions such as hypertension or to provide other health promoting activities.

What Is the Physiological Response of Animals to Feeding Hydrolysates?

With production animals, increased nitrogen absorption is a prerequisite to increased tissue mass and rate of muscle gain, which is important at all stages of growth (suckling, weaning, and post-weaning). However, very little research has been published that evaluates the influence of protein hydrolysates on animal performance in pigs and is currently limited to two experiments conducted on weaned piglets (1–23 days old; Pettigrew et al. 1977; Lindemann et al. 2000). Within the swine industry, nutritional support of the weanling is critical to improving production efficiency. Generally, these studies yield contradictory results in comparison, but the experimental animals were very different in weanling age. Therefore, direct comparison is not productive as weaned piglets at 1 and 2 days old is not a standard industry practice. Also, the diets did not contain the currently incorporated spray dried plasma protein in the formulation, which has been demonstrated to improve weaning performance (Kats et al. 1994). In summary, the link between the enhanced portal appearance of AAs with protein hydrolysates and animal performance evaluation in pigs has not been established.

In contrast to the minimal amount of performance research conducted in the pig, several studies have been conducted with rats to compare the effects of various sources of protein hydrolysates on weight gain, nitrogen digestibility, and net protein utilization (NPU) (Table 6). In general, healthy weanling rats appear to demonstrate no enhanced nitrogen digestibility or weight gain from feeding of hydrolysates versus intact protein, whereas whey and casein hydrolysates improved NPU by 13.6% (Boza et al. 1994) and 12.5% (Yamamoto et al. 1985) versus intact protein or FAAs, respectively. In addition, direct comparison of hydrolysates of different protein sources have demonstrated a 90% and 36% increase in weight gain (yolk versus soy; Gutierrez et al. 1998) after 7 and 28 days of feeding, which was in part a consequence of approximately 10% increase in feed intake. A 98% and 136% increase in NPU was also observed with milk hydrolysate versus bovine albumin hydrolysate over a 6 and 15 days period (Cezard et al. 1994). Interestingly, the milk hydrolysate contained 26% as di- and tripeptides with the remaining nitrogen as larger peptides (4–20 AA in length) compared to the albumin hydrolysate that contained 75% of nitrogen as di- and tri-peptides. This suggests that the constituent AA composition of the protein was more critical than peptide length in the nitrogen retention of these weanling rats.

Expression of coat color, coat health, optimal physical health, and longevity are measures of animal performance in the adult companion animal. These targeted goals are the underlying fundamentals in the nutritional support of the healthy adult companion animal in the maintenance state. Along with this is a growing need to better understand the AA requirements for cats and dogs, and between breeds (Hendricks 2003). Currently, direct evidence attributing any nutritional benefit for the use of hydrolysates in companion animal diets is largely undetermined, with the exception of the N-disappearance studies discussed above. The feeding of enteral diets containing protein as FAAs, peptides, and/or intact protein to malnourished veterinary patients is suggested in clinical veterinary practice (Marks 1998; Remillard et al. 2000), but to our knowledge no studies exist that compare hydrolysates to intact protein or FAA mixtures have been performed in companion animals to evaluate intestinal recovery, improved health or "performance". However, a single study did, at least, characterize portal AA profiles at various times in the postoperative conscious dog following jejunal infusion of a enteral hydrolysate (Bodoky et al. 1988).

Table 6 Affect of feeding AAs supplied as protein hydrolysate versus that by intact protein or FAAs on N digestibility, weight gain, and nitrogen retention by rats

Species	Feeding period, days	Protein type	Protein form[a]	Protein content	N-digest.[b]	Weight gain[c]	NPU[d]	Citation
Rat				%				
41 g	7	Yolk[g]	H versus I	10	0[e]	0	0	Gutierrez et al. 1998
50 g		Casein[h]	H versus I	12	0[f]		0	Boza et al. 1994
		Whey[i]	H versus I	12	0		+	
		Whey	H versus I	13.6	0[f]	0	0	Nakano et al. 1994b
80 g	28	Casein[j]	H versus I	15	0[e]	0	0	Yamamoto et al. 1985
	28	Casein[j]	H versus FAA	15	0	0	+	
100 g	10		H versus I			+		Zaloga et al. 1991
			H versus FAA			+++		

[a]Weight gain achieved by rats fed protein hydrolysate, relative to that by rats fed intact (I) or FAAs.
[b]Nitrogen digestibility.
[c]Relative weight gain at end of feeding period compared to start of feeding.
[d]Estimated apparent NPU = (Apparent Nitrogen Retained/Nitrogen Intake) × 100.
[e]Mean apparent nitrogen digestibility.
[f]Mean true nitrogen digestibility.
[g]Yolk: average peptide length was 2.6 AA residues.
[h]Casein: 72.1% 200–800 MW, 2.9% FAA.
[i]Whey: 44.5% 200–800 MW, 14.6% FAA, 44.3% 800–8,000 MW.
[j]Casein: 65% with average peptide as 5–6 AA in length, 35% as FAA.

What Is the Influence of Hydrolysates on the Malnourished Animal?

Although there is little evidence to support the use of diets formulated with protein hydrolysates to enhance whole-body performance in a healthy adult animal (Table 6), evidence is accumulating to indicate that protein hydrolysates do provide nutritional benefits to support the recovery of malnourished animals or animals requiring enteral feeding (Table 7). Common physiological stresses like weaning, injury, drug treatment, and/or sickness typically result in an adverse change in an animal's physiological status as a consequence of condition-induced malnourishment. During these conditions, diarrhea or decreased nutrient intake caused by complete (starvation) or partial (semi-starvation) cessation of food intake exacerbates nutrient deprivation, which compromises intestinal function and structure (Firmansah et al. 1989; Ferraris and Carey 2000). Recent reviews have discussed that malnutrition-induced villus atrophy can be defined as a decrease in villus height, decreased cell proliferation and migration rates, and increased rate of cell loss (Ferraris and Carey 2000), accompany an impairment of digestive and absorptive function in nursing and weaned pigs (Nunez et al. 1996; Pluske et al. 1996). The deprivation of nutrients leads to a whole-body catabolic state that negatively influences animal performance by reducing immunocompetence, decreasing tissue synthesis and repair, and/or alters intermediary drug metabolism (Remillard et al. 2000). Therefore, with the understanding that deterioration of the intestinal mucosa leads to a malnourished state, nutritional physiologists now recognize that specifically addressing the nutritional requirements of the intestine is a prerequisite to metabolic recovery and regrowth of catabolized tissue.

Formulating diets with predigested protein that contains small peptides has long been regarded as the ideal means of supplying absorbable N to meet the nitrogen requirements of the malnourished patient to surpass the loss of digestive capacity (Silk et al. 1980). This understanding was primarily based on the observations that the rate of AA absorption was faster when administered in the peptide form rather than as FAA (Silk et al. 1973a, b; Fairclough et al. 1975; Burston et al. 1980). More recently, feeding protein hydrolysate versus FAA-based diet is implicated in the direct recovery of intestinal tissue and contributes to peripheral tissue performance, which is discussed below.

To further study forms of dietary protein on recovery of the malnourished animal, starvation/refeeding models have been developed using adult or weanling rats deprived of food for 72 h, followed by feeding for various intervals. Several studies have been performed to evaluate the effect of feeding of dietary protein as intact, hydrolysates, or free AA mixtures on animal performance during recovery from a period of starvation (Table 7). In adult rats, refeeding of whey hydrolysates were directly compared to FAA mixtures over consecutive 24 h intervals through 96 h of refeeding (Poullain et al. 1989b) or after 72 h of refeeding (Boza et al. 2000). The refeeding of protein hydrolysates resulted in a greater weight gain and 30–250% increase in apparent NPU, which coincided with a 78–83% decrease in

Table 7 Affect of feeding AAs supplied as protein hydrolysate versus that by intact protein or FAAs on N digestibility, urinary N, weight gain, and nitrogen retention by 72-h starved rats[a]

Life stage, Weight	Protein source	Protein form	Refed, h	N- digest.[b]	Urine N loss	Weight gain[c]	NPU[d]	Citation
Adult	Whey	H versus I	Control[e]	0	---		+++	Poullain et al. 1989b
290 g			24	0	---		+++	
			48	0	---		+++	
			72	0	---		+++	
			96	0	---	++++	++	
		H versus FAA	Control[e]	0	---		++	
			24	0	---		++	
			48	0	---		++	
			72	0	---		+++	
			96	0	---	+++++	+++	
Adult	Whey	H versus FAA	72	0		+[f]	++[f]	Boza et al. 2000
200 g								
Weanling	Whey	H versus I	96	0	0	0	0	Boza et al. 1996
21 days old								
Weanling	Casein	H versus I	96			0		Boza et al. 1995a
21 days old								
Weanling	Whey	H versus I	48	0	-	0	+	Boza et al. 1995b
21 days old	Casein	H versus I	48	0	--[f]	0	++	

[a]Relative difference in animal performance parameters when refed with protein hydrolysate (H) versus intact protein (I) or FAAs (FAA). Relative changes: + = 10–25%; ++ = 25–50%; +++ = 50–100%, ++++ = 100–200%, +++++ = >200%; 0 = no difference (–10% to 10%); - = –10% to –25%; -- = –25% to –50%; --- = <–50%.

[b]Apparent nitrogen digestibility.

[c]Weight gain represents the relative difference in body weight of H versus I or FAA treatment animals, and from the end of a 72 h starvation period to end of refeeding period.

[d]Apparent Net Protein Utilization (NPU) = ((amount of nitrogen retained)/(amount of intake nitrogen) × 100).

[e]Control represents unstarved rats fed one of the three treatment diets for 96 h.

[f]$P < 0.05$.

urine N output versus feeding FAA. Villus height and crypt height were also higher with hydrolysate fed rats versus FAA through 72 h of refeeding (Poullain et al. 1989a). In weanling rats, results from studies that compared feeding protein as hydrolysate versus intact protein suggest that improved performance, as indicated by increased NPU and/or decreased urine-N output, is dependent on the period of refeeding following starvation (Boza et al. 1995a, b, 1996). Specifically, whole-body performance measurements indicate that protein hydrolysates support the metabolic recovery from starvation within 48 h of starvation, whereas the nutritional advantage of feeding hydrolysates over intact protein is diminished at 96 h following starvation.

Besides the 72-h starvation model (Table 7), N-absorption has been studied to evaluate hydrolysates as the optimal form of dietary protein during recovery from malnourished conditions resulting from infection and surgery. Malnourished mice subjected to schistosomiasis demonstrated 5–7% greater N-absorption with casein hydrolysate versus intact protein (Coutinho et al. 2002). Similarly after surgical procedures in humans, intestinal function and digestive capacity were compromised. The enteral delivery of nutrients with hydrolyzed protein demonstrated a greater N-digestibility and approximately 30% greater N-retention at day 6 (Ziegler et al. 1990), and increased appearance of peripheral plasma AA levels after 9 h, of feeding (Ziegler et al. 1998).

In addition to feeding protein hydrolysates, a Gln-containing peptide has demonstrated trophic effects on the small intestine. As briefly mentioned above, nearly 100% and 66% of dietary Glu and Gln, respectively, and approximately 30% of arterial Gln, are oxidized for energy by the intestinal mucosa (Wu 1998). In weanling pigs, supplementation with Gln reduces villus atrophy and improves weight gain (Wu et al. 1996). Also, in vitro research has demonstrated that "luminal fasted" human intestinal epithelial cells (Caco-2 cells) exhibit approximately a 40% decrease in intracellular Gln, Glu, and glutathione levels, along with a 20% decrease in the fractional protein synthesis rate and rise in transepithelial permeability, even when basolateral-facing membranes were exposed to nutrient rich conditions (simulating TPN; Le Bacquer et al. 2002). However, luminal or basolateral Gln or Glu supplementation restored the fractional protein synthesis rate and permeability.

Although we have described only two of many studies (Reeds et al. 1997; Ziegler et al. 2003) demonstrating the nutritional significance of Gln to the intestine, several recent studies using Ala-Gln have observed enhanced nutritional benefits compared to Ala and Gln supplied together in the free form. Specifically, mucosal wet weight, protein content, and villus height were significantly increased in rats administered Ala-Gln after experimental treatment to induce intestinal damage with a chemotheraputic drug (Satoh et al. 2003a) or an invasive surgical procedure involving resection of 80% of the proximal intestine (Satoh et al. 2003b). Additionally, in rats with cholera toxin-induced diarrhea, sodium and water secretion was decreased 128% and 95% with Ala-Gln as opposed to only a 36% and 60% reduction with Gln (Lima et al. 2002).

As a result of malnourishment, skeletal tissue catabolism involves release of Gln into the blood and a depletion of intracellular Gln levels, which is coupled to an inhibition of protein synthesis (Jepson et al. 1988). Malnourished animals fed hydrolysates have enhanced nitrogen retention (NPU) and greater weight gain, as compared to feeding FAA or intact protein. Accordingly, the reversal of the catabolic state may occur in response to reducing the intestinal requirement for energetic substrates like Gln supplied from the blood. This endogenous Gln "sparing" effect may result in greater Gln to stimulate insulin secretion (Li et al. 2004) and a faster accumulation of skeletal intracellular Gln levels. Recent research has demonstrated that both AAs and insulin are specifically involved in stimulating skeletal protein synthesis in 7 and 26 days old pigs (Davis et al. 2002), a developmental stage that is particularly prone to malnourishment after weaning. Consequently, the Gln-stimulated release of insulin, peripheral re-accumulation of Gln, and elevated plasma AA levels resulting from increased N-absorption all may be important factors for the stimulation of peripheral protein synthesis, and therefore the enhanced recovery of the malnourished state and regrowth of catabolized tissue.

Bioactive Dietary Peptides

The digestion of proteins to generate hydrolysates can yield peptides that have pharmacological roles, in addition to providing a readily absorbable nutrient source. Some of these bioactive peptides consist of only two or three AA residues, and have been identified as having therapeutic properties. Examples of di- and tripeptides that possess bioactive properties and are proven or putative PepT1 substrates have been collated in Table 8. These include β-alanyl-histidine (carnosine), peptides that contain Gly, Pro, or their derivatives (the GlyPro family; Maruyama et al. 1993; Samonina et al. 2002), and angiotensin I converting enzyme inhibitor peptides.

Table 8 Some examples of digestion-derived peptides with bioactive properties, which are known or putative substrates of PepT1-like

Protein source	Peptide	Citation
Animal tissue	Carnosine	Gariballa and Sinclair 2000
Wheat germ	Ile-Val-Tyr Val-Tyr	Matsui et al. 2000
Casein	Val-Pro-Pro Ile-Pro-Pro	Nakamura et al. 1995
Zein	Leu-Arg-Pro Leu-Ser-Pro Leu-Gln-Pro	Miyoshi et al. 1991
Collagen	Pro-Gly Gly-Pro Pro-Gly-Pro	Samonina et al. 2002
Collagen	Gly-Pro-Arg	Maruyama et al. 1993
Synthetic	Ala-Gln	Lima et al. 2002
Bonito muscle	Ile-Lys-Pro Ile-Trp Leu-Lys-Pro Leu-Tyr-Pro	Yokoyama et al. 1992
Sardine	Lys-Trp Met-Tyr	Matsufuji et al. 1994

Carnosine is an endogenously synthesized dipeptide present in the brain, cardiac muscle, kidney, and stomach tissues, and present in very high levels in avian flight and mammalian skeletal muscles (Gariballa and Sinclair 2000).

The role of carnosine in support of wound healing (Fitzpatrick and Fisher 1982; Nagai et al. 1986), vasodilation (Ririe et al. 2000), and as a biological buffer (Gariballa and Sinclair 2000) have been evaluated. The role of carnosine in improving wound healing is suggested to be partly related to its function as a dispensable pool of His (Nagai et al. 1986), with an enhanced response observed compared to administration of the corresponding FAAs (Roberts et al. 1998). Because carnosine is readily transported by PepT1 and relatively resistant to intestinal peptidase activity (Matthews et al. 1974), the use of dietary supplementation of carnosine as a therapeutic route of administration seems promising.

Some of the di- and tripeptides present in protein hydrolysates have been isolated and demonstrated to possess antihypertensive properties related to the inhibition of the angiotensin I-converting enzyme (Yamamoto 1997), thereby making their appearance, and potential therapeutic action, a consequence of PepT1-mediated absorption. Some examples of hypotensive peptides are Ile-Val-Tyr (Matsui et al. 2000), Val-Tyr (Matsui et al. 2000; Saiga et al. 2003), Val-Pro-Pro (Nakamura et al. 1995; Saiga et al. 2003), and Leu-Arg-Pro (Miyoshi et al. 1991). Furthermore, the potential roles of Pro-containing peptides, members of the GlyPro family (Pro-Gly, Gly-Pro, Pro-Gly-Pro, Hyp-Gly, Gly-Hyp, cycloPro-Gly), and Gly-Pro-Arg, in blood coagulation, platelet aggregation, and anti-ulcerogenic properties towards gastric mucosa (Maruyama et al. 1993; Samonina et al. 2002) have been evaluated.

Biochemical Characterization of PepT1 Activity

As discussed above, the elemental basis for why protein hydrolysates have demonstrated increased small intestinal N-disappearance, increased portal appearance of α-amino N, and increased whole-body performance during metabolic recovery from the malnourished state, is the existence of PepT1. PepT1 is believed to be responsible for absorption of thousands of potential combinations of AAs present as di- and tripeptides in the small intestinal lumen following protein digestion or feeding of a protein hydrolysate. In addition to having a nutritionally important role in the absorption of large portions and broad range of digestion-derived peptides present in the digesta, PepT1 has also been demonstrated to mediate the absorption of β-lactam antibiotics and ACE-inhibitors. In addition, PepT1 well recognizes at least several omega-amino fatty acids (Doring et al. 1998). For example, the affinity constants for PepT1 transport of 5-amino-pentanoic acid and 5-amino-4-oxo-pentanoic acid are 1.1 and 0.3 mM, respectively. In keeping with its ability to recognize a wide range of substrates, much of our current understanding about the molecular structure and functional activity of PepT1 has been supported by the pharmaceutical industry. The evaluation of PepT1 as a candidate drug delivery "system" has been extensively reviewed by others (Tsuji and Tamai 1996; Rubio-Aliaga and Daniel 2002; Brodin et al. 2002; Daniel 2004).

PepT1 is a member of the proton oligopeptide transporter (POT) family of integral membrane proteins capable of facilitating H^+-coupled peptide cotransport (Fei et al. 1998). Using the HUGO classification scheme nomenclature, PepT1 is listed as the first member of Solute Carrier Family 15 (i.e., SLC15.1). Expression cloning studies have identified cDNAs encoding PepT1-like transport activity from several different species, including rabbit (Fei et al. 1994), human (Liang et al. 1995), rat (Saito et al. 1995), mouse (Fei et al. 2000), chicken (Chen et al. 2002), and sheep (Pan et al. 2001). Characterization of the functional activity of PepT1 indicates that many PepT1 substrates have K_m values in the millimolar range, categorizing it as having low affinity substrate recognition. However, as noted above, the affinity of PepT1 for its substrates ranges over several orders of magnitude.

As illustrated in Fig. 1, the initial phase of transepithelial passage of di- and tripeptides requires an extracellular to intracellular pH gradient (Ganapathy et al. 1984). PepT1-mediated uptake of peptides is a H^+/peptide cotransport process (Daniel et al. 1991), with maximal functional activity was observed when the pH of the extracellular environment was 5.5 to 6.3. Cotransported H^+ is recycled back into the extracellular microenvironment in exchange for Na^+ by NHE3 (a H^+/Na^+ counterexchange transporter; Kennedy et al. 2002). Following intracellular flux of Na^+, cell membrane potential is maintained by the basolateral Na^+/K^+ ATPase responsible for the active transport of 3 Na^+ out of the cell in exchange for 2 K^+ taken into the cell. From an energy-expenditure perspective, PepT1 functions as a tertiary transporter, because the hydrolysis of ATP is performed by Na^+/K^+ ATPase (a primary transport activity) to pump H^+-exchanged Na^+ (a secondary transport activity) out of the cell. The importance of controlling the functional capacity of these primary and secondary transporters to support increased PepT1 functional capacity is discussed below in Section Regulation of PepT1 Functional Capacity.

Regulation of PepT1 Functional Capacity

Over the last half-century, many proteins present in the gastrointestinal tract critical to protein digestion and absorption of alpha-AA nitrogen have been identified. Yet, the underlying mechanisms responsible for responding to the likely dynamic range of intrinsic and extrinsic stimuli that regulate peptide and free AA absorption capacity have only recently begun to be revealed. As discussed above in some detail, it is clear that if the appropriate proportions of di- and tripeptides in protein hydrolysate are fed, then more AA N is absorbed as peptides then as FAA, and that PepT1 is most likely responsible for this phenomenon. Therefore, determining the cell-, tissue-, and whole body-physiological conditions that affect changes in the functional capacity of PepT1 become critical if we are to optimally match dietary nitrogen load with intestinal digestive and absorptive capacities.

Nutritional and Physiological and Conditions That Affect PepT1 Expression

As described earlier, it is estimated that 70–85% of total AA uptake from small intestinal digesta in healthy animals occurs in the form of peptides. Furthermore, feeding a peptide-containing protein hydrolysate following starvation provides a nutritional benefit to the malnourished animal (Table 7), which is at least partially attributed to PepT1 activity. A relationship exists between PepT1 expression and the malnourished state and/or mucosal damage, in that PepT1 expression in rats increases when they become malnourished as a result of starvation. However, the regulatory mechanisms activated by this physiological state have yet to be determined. In rats starved for 48 or 96 h, a significant decrease in mucosal weight and villus height is observed in all regions of the small intestine. At the same time, small intestinal PepT1 expression (mRNA and protein) and activity increases (Ogihara et al. 1999; Ihara et al. 2000; Naruhashi et al. 2002). Also, rats fasted for 24 h (Thamotharan et al. 1999a) or fed by total parenteral nutrition (TPN) (Ihara et al. 2000) experience loss of mucosal weight, which correlates with an up-regulation in small intestial PepT1 expression. The absence of luminal nutrients results in a depletion of intracellular stores of oxidizable AAs (Glu, Gln) that are not replenished by AAs available in the blood (basolateral-facing membrane), therefore the cell attempts to maximize uptake of α-amino nitrogen through the absorption of peptides by increasing the expression of small intestinal PepT1.

Similarly, certain drugs, bacterial intestinal infection, and/or intestinal surgery insults to intestinal mucosa also affect PepT1 expression, in a manner that appears specific to PepT1 and not other transporters of AAs. Specifically, the administration of cyclophosphamide (Satoh et al. 2003a) and 5-fluorouracil (Tanaka et al. 1998) (both known to cause small intestinal damage) to rats increased PepT1 mRNA 200% and 230%, respectively. In contrast to its up-regulation of PepT1 expression, 5-fluorouracil-induced mucosal damage resulted in a 50–85% decrease in mRNA for EAAC1 (high-affinity anionic transporter) and 4F2hc (the regulatory protein that co-localizes with either system $y^{+}L$ or $b^{0,+}$ transporters), whereas expression of ASCT2 (broad-spectrum neutral AA transporter) mRNA was insensitive to cyclophosphamide (Satoh et al. 2003a). During acute cryptosporidosis intestinal infection, PepT1 mRNA expression increased along the entire length of the small intestine (Barbot et al. 2003), whereas infection of the small intestinal mucosa by *N. brasiliensis* resulted in decreased jejunal, but not ileal, content of PepT1 mRNA (Sekikawa et al. 2003).

Small intestinal resection also typically results in the up-regulation of PepT1 mRNA expression. However, whether this is a response to trauma per se, or an attempt to recover lost PepT1 functional capacity, is unclear. For example, rat small intestinal PepT1 mRNA expression increased 150% by 5 days after removal of 80% of the proximal small intestine of rats (Satoh et al. 2003b). Similarly, removal of the human distal small intestine resulted in an five-fold increase of PepT1 mRNA content of the colonic epithelium (Ziegler et al. 2003).

Protein and Peptide Substrate Induced PepT1 Regulation

The regulation of PepT1 in response to various conditions of physiological stress underscores the nutritional importance of peptide absorption to meeting the nitrogen requirements of the animal when intestinal function is compromised. Importantly, PepT1 mRNA, protein, and functional capacity are all known to respond to dietary influences. Therefore, revealing the mechanisms by which dietary constituents may regulate PepT1-mediated absorption of peptides and beta-lactam antibiotics is of great interest to nutritionist and pharmacologists.

More thorough reviews of this topic exists elsewhere (Matthews 2000b; Meridith and Boyd 2000; Krehbiel and Matthews 2003). However, in general, increased levels of luminal protein and/or casein lead to an increase in PepT1 expression and/or activity. Ferraris and coworkers (1988), for example, examined the isonitrogenous feeding of casein in the form of intact, hydrolysate, or unbound (free) AAs on PepT1 functional capacity of the duodenum, jejunum, and ileum of mice. In this study the feeding of casein hydrolysate tended to increase PepT1 functional capacity, as compared to feeding FAAs. Similarly, endogenously expressed PepT1 activity of MDCK cells significantly increased when cultured in lactalbumin hydrolysate media, although this effect was not compared to a corresponding FAA media (Brandsch et al. 1995; Woods et al. 2001). Furthermore, some, but not all, peptides are capable of stimulating PepT1 mRNA, protein, and activity. In contrast to peptide substrate stimulation, PepT1 mRNA expression is down-regulated in rats administered FAA for 96 h (Ogihara et al. 1999). Whether this inhibitory effect results from specific or all AAs is unclear.

Cellular Control of PepT1 Functional Capacity Regulation

The above research has identified several physiological factors and substrates that affect the expression of PepT1. However, definition of the cellular pathways that mediate these responses are incomplete, even though variety of in vitro research has demonstrated that several hormones can elicit a regulatory response (Table 9). A variety of hormones have been evaluated to determine the affect on PepT1 expression using Caco-2 cells (Table 9; Adibi 2003; Nielsen and Brodin 2003). Insulin is known to be a postprandial metabolic regulator of plasma glucose and protein synthesis, and its secretion is stimulated by various nutrients and meal-induced autonomic nervous activation (Wollheim and Biden 1986; Benthem et al. 2000). Therefore, the postprandial role of insulin in influencing peptide absorption has been investigated as a regulator of PepT1 activity. Short-term exposure (2 h) of Caco-2 cells to insulin increased PepT1 transport capacity (Thamotharan et al. 1999b). A critical observation from this study was that insulin did not stimulate de novo synthesis of PepT1 mRNA or protein, but rather activation of sub-membrane vesicle migration and insertion of preformed PepT1 protein into the apical membrane.

Table 9 Hormonal regulation of PepT1 expression by Caco-2 cells[a, b]

Horm	Recept.	Experimental condition	Membrane exposure	Trt time	mRNA	Protein	MP[c]	Activity Substrate	K_t (mM)	V_{max}	Citation
Insulin	Ins-R	5 nM	Monolayer	2 h	1	1	1	GlyGln	1.36 ± 0.33	6.3 ± 0.5^{x}	Thamotharan et al. 1999a, b
		Control			1	1	2x^{x}	GlyGln	1.49 ± 0.55	3.5 ± 0.6	
		5 nM	Monolayer	2 h	1		1	CFX	5.37 ± 0.33	32.2 ± 3.9^{x}	Watanabe et al. 2003
		Control			1		1.6x^{x}	CFX	5.12 ± 0.43	14.8 ± 1.8	
		5 nM	AP	2 h				CFX		2.0^{x}	
		5 nM	BL	2 h				CFX		1.0	
		Control						CFX		1.0	
		50 nM	AP	15 days				CFX		12.3 ± 1.1^{x}	
		50 nM	BL	15 days	1			CFX		7.7 ± 0.5^{x}	
		Control			1			CFX		5.1 ± 0.4	
		50 ng*mL^{-1}	BL	1 h	1	1		GlySar	1.03 ± 0.08	2.2 ± 0.1	Nielsen et al. 2003
		Control	BL		1	1		GlySar	0.98 ± 0.11	1.9 ± 0.1	
Leptin	Ob-Rb	2 nM	AP	1 h	1		1.4x^{x}	GlySar	0.31 ± 0.04	6.5 ± 0.3^{y}	Buyse et al. 2001
		Control			1		1	GlySar	0.30 ± 0.03	4.3 ± 0.1	
		100 nM	AP	1 h				CFX		9.0^{y}	
		Control						CFX		4.0	
EGF	EGFR	200 ng*mL^{-1}	BL	1 h	1	1		GlySar	1.11 ± 0.05	2.8 ± 0.1	Nielsen et al. 2003
		Control			1	1		GlySar	0.98 ± 0.11	1.9 ± 0.1	
		5 ng*mL^{-1}	BL	>26 days	0.65x^{e}	0.65x^{e}		GlySar	0.57 ± 0.20	1.1 ± 0.1	Nielsen et al. 2001
		Control			1^{e}	1^{e}		GlySar	0.66 ± 0.30	2.6 ± 0.4^{x}	

IFN-γ	nd[d]	100 IU*mL^{-1}	BL	48 h	1	1	GlySar		0.56 ± 0.01[x]	Buyse et al. 2003
		Control			1	1	GlySar		0.35 ± 0.03	
rhGH	nd[d]	34 nM	AP	4 days			CFX		3.5	Sun et al. 2003a
		Control					CFX		1.8	
		34 nM-anoxia	AP	4 days	1.3x		CFX		2.3	
		Control-anoxia[f]			1		CFX		1.3	
VIP	$VPAC_1$	5 nM					GlySar		0.35	Anderson et al. 2003
		Control					GlySar		0.65[y]	
T_3	nd	100 nM	Monolayer	4 days	0.25x	0.75x	GlySar	1.02 ± 0.26	5.2 ± 1.0	Ashida et al. 2001
	α_2-receptor	Clonidine 10μM	BL	15 min			CFX		92.8 ± 13.4	Berlioz et al. 2000
		Control					CFX		54.4 ± 9.1[x]	
	σ-receptor	Pentazocine 1μM	Monolayer	24 h	2x[y]		GlySar		10.3 ± 0.7	Fujita et al. 1999
		Control			1		GlySar		20.9 ± 1.1	

[a]Hormone treatment time and membrane exposure are indicated, whereas controls lacked hormone treatment. Controls are listed below the corresponding hormone treatment.

[b]PepT1 mRNA and protein expression and apical membrane protein content are listed as X-fold change versus corresponding control: [x]$P < 0.05$, [y]$P < 0.01$.

[c]MP: apical membrane protein content.

[d]nd: not determined.

[e]Protein and mRNA levels are representative of EGF treatment in culture media for 24 days.

[f]Caco-2 cells were subjected to a 90-min period of anoxia, followed by a 30-min period of reoxygenation.

Subsequently, this observation was confirmed (Nielsen et al. 2003; Watanabe et al. 2003). However, these subsequent studies led to contradictory findings for determining the insulin-responsive plasma membrane. Membrane characterization of Caco-2 and canine intestinal epithelial cells has demonstrated that the insulin receptor is primarily localized to the basolateral membrane (Gingerich et al. 1987; MacDonald et al. 1993). Therefore, apical membrane stimulation of PepT1 activity by insulin was suggested to result from differential membrane expression of insulin receptor under slightly different culture conditions (Watanabe et al. 2003). Regardless, it now appears that an emerging theme for how insulin regulates short-term transport capacity of blood glucose and small peptides is through the recruitment of submembrane vesicles of preformed transporters.

Insulin is not the only hormone associated with whole-body nutritional status that is capable of stimulating PepT1 function activity. Leptin is secreted from the stomach into the blood and gastric lumen following feeding or exogenous CCK-8 administration in rats (Bado et al. 1998), or following infusion of pentagastrin or secretin in humans (Sobhani et al. 2000). Similar to insulin, short-term (60 min) exposure of Caco-2 cells to leptin also stimulates increased PepT1 activity by incorporation of submembrane vesicles containing PepT1 into the plasma membrane (Buyse et al. 2001). Unlike insulin, however, leptin-stimulated PepT1 activity occurs through the apical membrane, not basolateral exposure (Buyse et al. 2001). This understanding is consistent with the localization of the Ob-Rb leptin receptor in the apical membrane of Caco-2 cells and rat small intestinal epithelia (Buyse et al. 2001). The Ob-Rb leptin receptor mediates the activation of phosphoinositide 3-kinase (PI3-K) through activation of insulin receptor substrate 1 and 2, albeit at a lesser sensitivity than insulin itself (Bjorbaek et al. 1997; Kim et al. 2000). Therefore, the PI3-K pathway appears to be stimulated by leptin and insulin, and correspondingly up-regulation of PepT1 activity through recruitment of PepT1-containing submembrane vesicles might be controlled by both luminal and blood regulatory factors.

Trophic hormones not stimulated by postprandial events, but which affect PepT1 expression and functional activity are known. Specifically, epidermal growth factor (EGF) binding to the tyrosine kinase epidermal growth factor receptor (EGFR) stimulates cell proliferation, differentiation, migration, survival, and adhesion (Janmaat and Giaccone 2003) events in a variety of cell types, including gut mucosa (Sarosiek et al. 1988, 1991; Fisher and Lakshmanan 1990). Synthesis of EGF occurs in various tissues, including the mammary gland (Kajikawa et al. 1991) and the submandibular salivary gland (Abdollahi and Simaiee 2003). Consequently, salivary and mammary glands are likely gut lumen sources of EGF for activation of EGF receptor (EGFR), which is localized in the apical membrane of gastrointestinal tract epithelia (Schweiger et al. 2003). In contrast, peripheral tissues such as the pancreas, liver, muscle, kidney, thyroid gland, and anterior pituitary (Vaughan et al. 1991; Kajikawa et al. 1991; Fan and Childs 1995) express EGF mRNA, thus making them likely suppliers of plasma EGF. Concomitant with identification of EGF sources, the pattern of EGFR expression has been evaluated. Studies of intestinal tissues indicate that membrane localization and cell-type expression differs among ontogenic stages and between species (Kelly et al. 1992; Playford et al. 1996; Schweiger et al. 2003). In Caco-2 cells, EGFR localizes to both the apical and basolateral membranes (Bishop and Wen 1994).

Although Caco-2 cells express EGFR in both apical and basolateral membranes, EGF treatment has been shown to affect PepT1 expression only through basolateral membrane exposure. However, seemingly contradictory results have been observed regarding PepT1 mRNA and functional activity in a variety of studies. Long-term (15 days) basolateral (but not apical) exposure of EGF strongly inhibited PepT1 activity, resulting in a 60% decrease in the V_{max} of GlySar uptake and a 35% decrease in both PepT1 mRNA and protein content (Nielsen et al. 2001). In contrast, short-term (60 min) basolateral (but not apical) exposure to EGF caused an increase in PepT1 activity, but without alteration of PepT1 mRNA content (Nielsen et al. 2003). To evaluate whether the mechanisms of action was through recruitment of submembrane vesicles, and therefore similar to the mode of insulin and leptin stimulation of PepT1 activity, the microtubule network of EGF-treated Caco-2 cells was disrupted and PepT1 functional activity measured. However, this treatment did not statistically reduce the stimulatory effect of EGF. Thus, the mechanism by which short-term exposure of EGF stimulates PepT1 activity remains unknown.

A combination of EGF plus GH results in a two-fold increase in jejunal PepT1 mRNA (Avissar et al. 2001). Whether this increase in mRNA level corresponds to an increase in PepT1 functional activity was not reported. Further evidence suggesting that GH influences PepT1 has been obtained through in vivo and in vitro experimentation (Sun et al. 2003a, b). For example, PepT1 transport capacity was suggested to be positively influenced by recombinant human growth hormone (rhGH) administered to rats with 3° burns and in Caco-2 cells, which resulted in a 1.3-fold increase in PepT1 mRNA in Caco-2 cells. However, rhGH was exposed to the apical membrane of the Caco-2 cells, which is in contrast to the parenteral administration of EGF plus GH after intestinal resectioning (Avissar et al. 2001) or rhGH in burned rats (Sun et al. 2003b). Also, cephalexin uptake studies in rhGH exposed Caco-2 cells were performed at pH 7.0, which would not discriminate from H^+-dependent and non-mediated absorption of cephalexin. Therefore, further work is required to characterize the potential influence of rhGH on PepT1 expression and activity.

Studies designed to evaluate the potential interaction of PepT1 and immune responses have shown that PepT1 mediates the absorption of a peptide (formyl-Met-Leu-Phe) of microbial origin, known to stimulate intestinal inflammatory response and that interferon-gamma (IFN-γ) stimulates PepT1 activity by 60% in Caco-2 cells (Buyse et al. 2003). Furthermore, the increase in PepT1 activity was not paralleled by increased PepT1 mRNA or protein, but was concomitant with an increase in intracellular pH. This observation appears to be paradoxical because Na^+/H^+ exchangers are known to be integral to PepT1 function by exchanging the peptide-coupled symport of H^+ for extracellular Na^+ (Fig. 1) and because IFN-γ has been reported to down-regulate the NHE2 and NHE3 Na^+/H^+ exchangers in Caco-2 cells (Rocha et al. 2003). Therefore, the increased functional activity of PepT1 was unexpected. However, it is thought that increases in intracellular pH are associated with Cl^-/HCO_3^- basolateral membrane exchanger activity (Busche et al. 1993). Through this mechanism, IFN-γ may act to disassociate H^+-dependent mediated peptide transport from the corresponding secondary transport process of Na^+/H^+ exchanger proteins leading to the inhibition of ion transport and contributing to the development of diarrhea, while also maintaining delivery of antigenic peptides.

PepT1 expression and activity can be negatively influenced by hormones other than long-term exposure of EGF, including thyroid hormone (T_3) and vasoactive intestinal peptide (VIP). Treatment of Caco-2 cells for 4 days with 100 nM T_3 caused a decrease in PepT1 expression (mRNA and protein) and activity (Ashida et al. 2001), but the mechanism is unknown. In contrast, VIP and pituitary adenylate cyclase-activating polypeptide elicit an inhibitory effect on the uptake of GlySar by Caco-2 cells through activation of the $VPAC_1$ receptor (Anderson et al. 2003). VIP binding to the $VPAC_1$ receptor triggers the activation of adenylate cyclase for synthesis and rise in cAMP for the consequential activation of protein kinase A (PKA). Because elevated cAMP levels are known to inhibit PepT1 activity (Muller et al. 1996), the human colonocyte cell line, Caco-2, was genetically engineered to express the α_2-receptor to more directly evaluate how cAMP affects PepT1 expression. The α_2-receptor is endogenously expressed in normal intestinal cells and is coupled to the inhibitory G-proteins, G_{i2} and G_{i3}, which act to inhibit cAMP production. Treatment of the modified Caco-2 cells with an α_2-receptor agonist (clonidine) stimulated a two-fold increase in PepT1 transport activity, through clonidine inhibition of forskolin-stimulated cAMP production (Berlioz et al. 2000), thus confirming the inhibitory role of cAMP on PepT1 activity.

To confirm that PKA was involved in the observed inhibition of PepT1, treatment of Caco-2 cells with a PKA inhibitor (H-89) was evaluated and found to block the inhibition of GlySar uptake. Furthermore, GlySar uptake can also be inhibited by a compound (S1611) that specifically inactivates the regulatable apical Na^+/H^+ exchanger (NHE3). Therefore, VIP binding to the $VPAC_1$ receptor triggers activation of PKA, which leads to the inactivation of lumen-facing NHE3, ultimately preventing acidification of the extracellular microenvironment and reducing PepT1 uptake capacity. Thus, VIP appears to inhibit PepT1 functional capacity by reducing its requisite H^+-driving force. This hypothesis is consistent with the "anti-absorptive" actions of VIP (Carter et al. 1978).

Obviously, much work has been conducted to elucidate putative pathways for leptin, insulin, GH, T3, EGF, VIP, and/or IFN-γ regulation of PepT1 functional capacity. To summarize these seminal efforts and to highlight where control of these pathways might converge, our working model of prominent putative regulatory pathways for PepT1 and functionally-associated proteins is presented in Fig. 2. In this model, either with plasma insulin activation of the basolateral membrane-bound insulin receptor, or with luminal leptin activation of apical membrane-bound leptin receptor (Ob-Rb) the migration of submembrane vesicle-bound PepT1 molecules to the apical membrane is initiated. VIP binds the VPAC1 receptor, thereby stimulating cAMP synthesis by adenylate cyclase and subsequent activation of protein kinase A. Activated PKA inhibits the activity of the NHE3, resulting in reduced extracellular to intracellular H^+ gradient that drives optimal PepT1 activity.

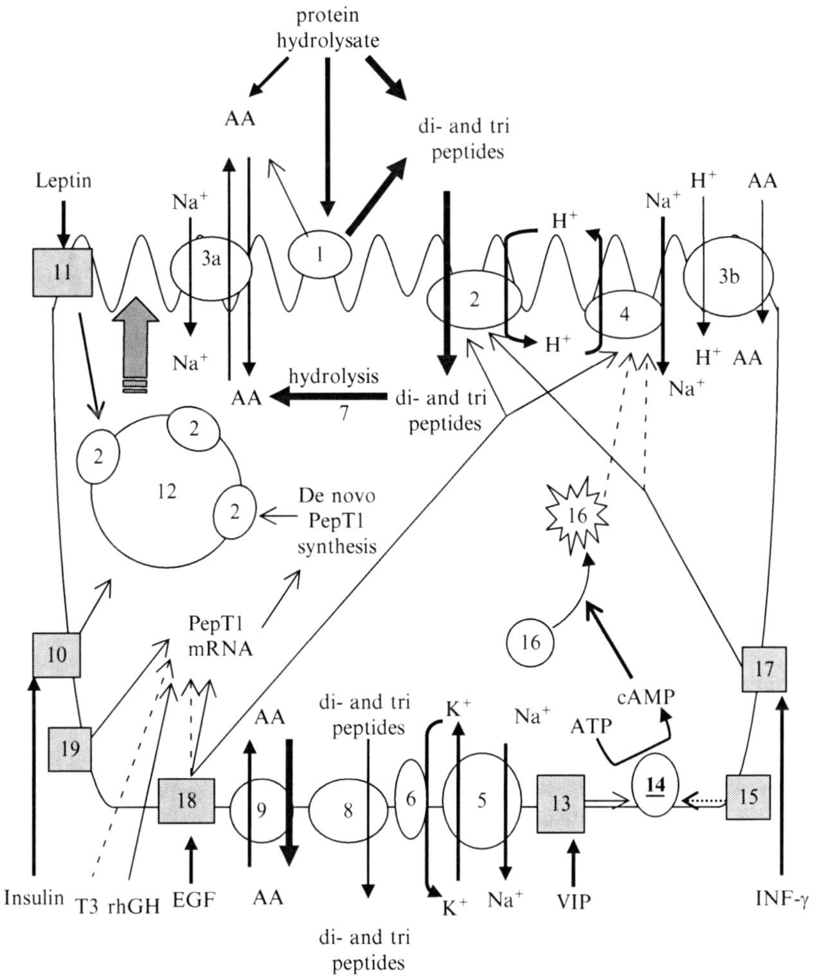

Fig. 2 Model for regulation of PepT1 functional capacity. Nutrient flux or receptor binding (closed-head arrows) and activation (solid line) or inhibition (broken line) of pathway or protein (open-head arrows) are indicated. Model function are described in text. 1, membrane-bound peptidases; 2, PepT1; 3a and 3b, Na^+-dependent and H^+-dependent AA transporters, respectively; 4, NHE3; 5, Na^+/K^+ ATPase; 6, K^+ channel; 7, intracellular hydrolysis; 8, H^+-independent peptide transporter; 9, basolateral AA transporters; 10, insulin receptor; 11, leptin receptor; 12, vesicle-bound PepT1; 13, VPAC1 receptor; 14, adenylate cyclase; 15, α_2-receptor; 16, PKA; 17, INF-γ receptor; 18, EGF receptor; 19, sigma receptor

In contrast, activation of the α_2-receptor maintains the H^+ gradient by inactivating adenylate cyclase. On the other hand, INF-γ receptor activation putatively stimulates PepT1 function while inhibiting that of NHE3. Short-term binding of the EGF receptor leads to activation of PepT1 and NHE3, whereas long-term exposure decreases in PepT1 gene transcription. Activation of the sigma receptor leads to an increase in PepT1 mRNA, whereas rhGH and T_3 reduce PepT1 mRNA content.

Summary and Conclusions

To increase the efficiency of protein assimilation, the challenge to nutritionists is to match gastrointestinal tract load to digestive and absorptive capacities. Supplying dietary protein in the form of milk or egg protein hydrolysates that contains small peptides bypasses luminal digestion events and increases appearance of α-amino nitrogen in portal and peripheral blood in healthy rats, pigs, dogs, and humans, when compared to the delivery of the same protein as FAAs. This difference is maximized when the concentration of peptides approaches typical K_m values of the intestinal peptide transporter (PepT1), as opposed to typical K_m values for AA transporters. Furthermore, appearance of specific AA in portal and peripheral blood is influenced by dietary peptide supply. Thus, peptide supplementation of the diet may be an effective strategy to deliver desired AAs to support intestinal and peripheral tissues metabolism and, thereby, improve net protein utilization and N retention. However, the evidence is incomplete as to whether this strategy will work for healthy animals (including humans), as it seems to for metabolically-stressed animals.

PepT1 activity is the known transporter for absorption of small peptides from digesta by the gut epithelium that functions as a high-capacity transporter, low-affinity transporter, displaying a large substrate recognition profile for di- and tripeptides and certain classes of pharmaceutical compounds, such as β-lactam antibiotics. PepT1 functional capacity is up-regulated by a variety of physiological stimuli, including substrate load and disease state. Although the cellular mechanisms by which this regulation occurs have not been elaborated, leptin and insulin (and/or EGF) appear to be plausible candidates for hormonal control of PepT1 functional capacity through luminal- and blood-interactions, respectively.

Future Developments

The most immediate application of the reviewed biological knowledge likely exists (and in some cases is occurring) is the use of protein hydrolysates in the formulation of diets of metabolically-stressed animals. That is, peptide-supplied AAs may be especially beneficial for animals when the demand for AAs of a particular physiological state exceeds the in vivo digestion capacity for feedstuff proteins. Because of the added cost of producing the peptides, the economical use may be restricted to periods of acute stress, such as during weaning and shipping of young mammals. In addition, use of peptide-containing enteral formulas for human and veterinary clinical cases following illness or surgery may be a viable alternative to formulas that currently contain intact protein or FAAs.

In the future, however, with the maturation of the biotechnology industry, the large-scale production of transgenically-derived animal proteins may become economically feasible, especially if the cost-benefit market analysis includes the potential for loss of market share due to psychological and/or physiological-based fears. For example, the production of milk and/or egg proteins by transgenic plants, and their subsequent in vitro digestion to small peptides, would allow the biological value of these proteins to be realized but without associated "risk" of prion and viral diseases that accompany the use of endogenously derived animal proteins.

References

Abdollahi M, Simaiee B (2003) Stimulation by theophylline and sildenafil of rat submandibular secretion of protein, epidermal growth factor and flow rate. Pharmacol Res 48:445–449

Adibi S (2003) Regulation of expression of the intestinal oligopeptide transporter (Pept-1) in health and disease. Am J Physiol Gastrointest Liver Physiol 285:G779–G788

Anderson CMH, Mendoza ME, Kennedy DJ, Raldua D, Thwaites DT (2003) Inhibition of intestinal dipeptide transport by the neuropeptide VIP is an anti-absorptive effect via the $VPAC_1$ receptor in a human enterocyte-like cell line (Caco-2). Br J Pharmacol 138:564–573

Ashida K, Katsura T, Motohashi H, Saito H, Inui KI (2001) Thyroid hormone regulates the activity and expression of the peptide transporter PEPT1 in Caco-2 cells. Am J Physiol Gastrointest Liver Physiol 282:G617–G623

Avissar NE, Ziegler TR, Wang HT, Gu LH, Miller JH, Iannoli P, Leibach FH, Ganapathy V, Sax HC (2001) Growth factors regulation of rabbit sodium-dependent neutral AA transporter ATB0 and oligopeptide transporter 1 mRNAs expression after enteretomy. Journal of Parenteral and Enteral Nutrition 25:65–72

Bado A, Levasseur S, Attoub S, Kermorgant S, Laigneau JP, Bortoluzzi MN, Moizo L, Lehy T, Guerre-Millo M, Le Marchand-Brustel Y, Lewin MJ (1998) The stomach is a source of leptin. Nature 394:790–793

Barbot L, Windsor E, Rome S, Tricottet V, Reynes M, Topouchian A et al (2003) Intestinal peptide transporter PepT1 is over-expressed during acute cryptosporidiosis in suckling rats as a result of both malnutrition and experimental parasite infection. Parasitol Res 89:364–370

Baro L, Guadix EM, Martinez-Augustin O, Boza JJ, Gil A (1995) Serum AA concentrations in growing rats fed intact protein versus enzymatic protein hydrolysate-based diets. Biol Neonate 68:55–61

Benthem L, Mundinger TO, Taborsky GJ Jr (2000) Meal-induced insulin secretion in dogs is mediated by both branches of the autonomic nervous system. Am J Physiol Endocrinol Metab 278:E603–E610

Berlioz F, Maoret J, Paris H, Laburthe M, Farinotti R (2000) α_2-Adrenergic receptors stimulate oligopeptide transport in a human intestinal cell line. J Pharmacol Exp Ther 294:466–472

Bishop WP, Wen JT (1994) Regulation of Caco-2 cell proliferation by basolateral membrane epidermal growth factor receptors. Am J Physiol 267:G892–G900

Bjorbaek C, Uotani S, da Silva B, Flier JS (1997) Divergent signaling capacities of the long and short isoforms of the leptin receptor. J Biol Chem 272:32686–32695

Bodoky A, Heberer M, Landmann J, Fricker R, Behrens D, Steinhardt J, Harder F (1988) Absorption of protein in the early postoperative period in chronic conscious dogs. Experimentia 44:158–161

Boza JJ, Jiminez J, Martinez O, Suarez MD, Gil A (1994) Nutritional values and antigenicity of two milk protein hydrolysates in rats and guinea pigs. Am Inst Nutr 1978–1986

Boza JJ, Martinez O, Baro L, Suarez MD, Gil A (1995a) Influence of casein and casein hydrolysate diets on nutritional recovery of starved rats. Journal of Parenteral and Enteral Nutrition 19:2126–2221

Boza JJ, Martinez-Augustin O, Baro L, Suarez MD, Gil A (1995b) Protein versus enzymic protein hydrolysates, Nitrogen utilization in starved rats. Br J Nutr 73:65–71

Boza JJ, Jiminez J, Baro B, Martinez O, Suarez MD, Gil A (1996) Effects of native and hydrolyzed whey protein on intestinal repair of severely starved rats at weaning. J Pediatr Gastroenterol Nutr 22:186–193

Boza JJ, Moennoz D, Vuichoud J, Jarret A, Gaudard-de-Weck D, Ballevre O (2000) Protein hydrolysate versus free AA-based diets on the nutritional recovery of the starved rat. Eur J Nutr 39:237–243

Brandsch M, Ganapathy V, Leibach F (1995) H+-peptide cotransport in Madin-Darby canine kidney cells: expression and calmodulin-dependent regulation. Am J Physiol 268:F391–F397

Brodin B, Nielsen C, Steffansen B, Frokjaer S (2002) Transport of Peptidomimetic drugs by the intestinal di/tri-peptide transporter, PepT1. Pharmacol Toxicol 90:285–296

Buraczewska L (1981) Absorption of AAs in different parts of the small intestine in growing pigs. iii. absorption of constituents of protein hydrolysates. Acta Physiol Pol 32:569–584

Burrin DG, Reeds PJ (1997) Alternative fuels in the gastrointestinal tract. Curr Opin Gastroenterol 13:165–170

Burston D, Taylor E, Matthews D (1980) Kinetics of uptake of lysine and lysyl-lysine by hamster jejunum in vitro. Clin Sci 59:285–287

Busche RJ, Jeromin A, Von Engelhard W, Rechkemmer G (1993) Basolateral mechanisms of intracellular pH regulation in the colonic epithelial cell line HT29 clone 19A. Eur J Physiol 425:219–224

Buyse M, Berlioz F, Guilmeau S, Tsocas A, Voisin T, Peranzi G et al (2001) PepT1-mediated epithelial transport of dipeptides and cephalexin is enhanced by luminal leptin in the small intestine. J Clin Investig 108:1483–1494

Buyse M, Charrier L, Sitaraman S, Gewirtz A, Merlin D (2003) Interferon-γ increases hPepT1-mediated uptake of di-tripeptides including the bacterial tripeptide fMLP in polarized intestinal epithlia. Am J Pathol 163:1969–1977

Carter RF, Bitar KN, Zfass AM, Makhlouf GM (1978) Inhibition of VIP-stimulated intestinal secretion and cyclic AMP production by somatostatin in the rat. Gastroenterology 74:726–730

Cezard JP, Tran TA, Macry J, Zarrabian S, Roger L, Bressolier P, Julien R, Mendy F, Kahn JM (1994) Effects of two protein hydrolysates on growth, nitrogen balance and small intestine adaptation in growing rats. Biol Neonate 65:60–67

Chen H, Pan Y, Wong E, Webb K Jr (2002) Characterization and regulation of a cloned ovine gastrointestinal peptide transporter (oPepT1) expressed in a mammalian cell line. J Nutr 132:38–42

Coutinho E, Ferreira H, Assuncao M, Carvalho S, Oliveira S, Francelino A (2002) The use of protein hydrolysate improves the protein intestinal absorption in undernourished mice infected with Schistosoma mansoni. Rev Soc Bras Med Trop 35:585–590

Daniel H (2004) Molecular and integrative physiology of intestinal peptide transport. Annu Rev Physiol 66:361–384

Daniel H, Morse EL, Adibi SA (1991) The high and low affinity transport systems for dipeptides in kidney brush border membrane respond differently to alteration in pH gradient and membrane potential. J Biol Chem 266:19917–19924

Davis TA, Fiorotto ML, Burrin DG, Vann RC, Reeds PJ, Nguyen HV, Beckett PR, Bush JA (2002) Acute IGF-I infusion stimulates protein synthesis in skeletal muscle and other tissues of neonatal pigs. Am J Physiol Endocrinol Metab 283:E638–E647

Defilippi C, Cumsille F (2001) Small-intestine absorption during continuous intraduodenal infusion of nutrients in dogs. Nutrition 17:254–258

Doring F, Will J, Amasheh S, Clauss W, Ahlbrecht H, Daniel H (1998) Minimal molecular determinants of substrates for recognition by the intestinal peptide transporter. J Biol Chem 4:23211–23218

Fairclough P, Silk D, Clark M, Dawson A (1975) New evidence for intact di- and tripeptide absorption. Gut 16:843, Abstract

Fan X, Childs GV (1995) Epidermal growth factor and transforming growth factor-alpha messenger ribonucleic acids and their receptors in the rat anterior pituitary: localization and regulation. Endocrinology 136:2284–2293
Fei YJ, Kanai Y, Nussberger S, Ganapathy V, Leibach FH et al (1994) Expression cloning of a mammalian proton-coupled oligopeptide transporter. Nature 7:563–566
Fei YJ, Ganapathy V, Leibach FH (1998) Molecular and structural features of the proton-coupled oligopeptide transporter superfamily. Prog Nucleic Acid Res Mol Biol 58:239–261
Fei YJ, Sugawara M, Liu JC, Li HW, Ganapathy V, Ganapathy ME, Leibach FH (2000) CDNA structure, genomic organization, and promoter analysis of the mouse intestinal peptide transporter PepT1. Int J Biochem Biophys 1492:145–154
Ferraris R, Carey H (2000) Intestinal transport during fasting and malnutrition. Annu Rev Nutr 20:195–219
Ferraris RP, Diamond J, Kwan WW (1988) Dietary regulation of intestinal transport of the dipeptide carnosine. Am J Physiol 255:G143–G150
Firmansah A, Suwandito L, Penn D, Lebenthal E (1989) Biochemical and morphological changes in the digestive tracts of rats after prenatal and postnatal malnutrition. Am J Clin Nutr 50:261–268
Fisher DA, Lakshmanan J (1990) Metabolism and effects of epidermal growth factor and related growth factors in mammals. Endocr Rev 11:418–442
Fitzpatrick DW, Fisher H (1982) Carnosine, histidine, and wound healing. Surgery 91:56–60
Fujita T, Majikawa Y, Umehisa S, Okada N, Yamamoto A, Ganapathy V, Leibach F (1999) σ-Receptor ligand-induced up-regulation of the H+/peptide transporter PEPT1 in the human intestinal cell line Caco-2. Biochem Biophys Res Commun 261:242–246
Ganapathy V, Burckhardt G, Leibach FH (1984) Caharacteristics of glycylsarcosine transport in rabbit intestinal brush-border membrane vesicles. J Biol Chem 259:8954–8959
Ganapathy V, Brandsch M, Leibach F (1994) Intesinal transport of AAs and peptides. In: Johnson LR (ed) Physiology of the gastrointestinal tract, 3rd edn. Raven Press, New York, pp 1773–1794
Gardner ML (1982) Absorption of intact peptides: studies on transport of protein digests and dipeptides across rat small intestine in vitro. Q J Exp Physiol 67:629–637
Gardner ML (1984) Intestinal assimilation of intact peptides and proteins from the diet-a neglected field? Biol Rev 59:289–331
Gardner MLG (1994) Absorption of intact proteins and peptides. In: Johnson LR (ed) Physiology of the gastrointestinal tract, 3rd edn. Raven Press, New York, pp 1795–1820
Gardner ML, Lindblad BS, Burston D, Matthews DM (1983) Trans-mucosal passage of intact peptides in the guinea-pig small intestine in vivo: a re-appraisal. Clin Sci 64(4):433–439
Gariballa SE, Sinclair AJ (2000) Carnosine: physiological properties and therapeutic potential. Age Ageing 29:207–210
Gingerich RL, Gilbert WR, Comens P, Gavin JR 3rd (1987) Identification and characterization of insulin receptors in basolateral membranes of dog intestinal mucosa. Diabetes 36:1124–1129
Giusi-Perier A, Fiszlewicz M, Rerat A (1989) Influence of diet composition on intestinal volatile fatty acid and nutrient absorption in unanesthetized pigs. J Anim Sci 67:386–402
Grimble GK, Rees RG, Keohane PP, Cartwright T, Desreumaux M, Silk DB (1987a) Effect of Peptide chain length on absorption of egg protein hydrolysates in the normal human jejunum. Gastroenterology 92:136–142
Grimble GK, Keohane PP, Higgins BE, Keminski MV Jr, Silk DB (1987b) Effect of peptide chain length on AA and nitrogen absorption from two lactalbumin hydrolysates in the normal human jejunum. Clin Sci 71:65–69
Gutierrez MA, Mitsuya T, Hatta H, Koketsu M, Kobayashi R, Juneja LR, Kim M (1998) Comparison of egg-yolk protein hydrolysate and soyabean protein hydrolysate in terms of nitrogen utilization. Br J Nutr 80:477–484
Han XT, Xue B, Du JZ, Hu LH (2001) Net fluxes of peptide and AA across mesenteric-drained and portal-drained viscera of yak cows fed a straw-concentrate diet at maintenance level. J Agric Sci 136:119–127
Hara H, Funabiki R, Iwata M, Yamazaki K (1984) Portal absorption of small peptides in rats under unrestrained conditions. J Nutr 114:1122–1129

Hegarty JE, Moriarty PD, Fairclough KJ, Kelly MJ, Clark ML (1982a) Effects of concentration on in vivo absorption of a peptide containing protein hydrolysate. Gut 23:304–309

Hegarty JE, Fairclough PD, Moriarty KJ, Clark ML, Kelly MJ, Dawson AM (1982b) Comparison of plasma and intraluminal AA profiles in man after meals containing a protein hydrolysate and equivalent AA mixture. Gut 23:670–674

Heger J (2003) Essential to non-essential AA ratios. In: D'Mello JPF (ed) AAs in animal nutrition. CAB International, Wallingford, UK, pp 103–124

Hendricks W (2003) Canine and feline AA requirements for different physiological functions. In: D'Mello JPF (ed) AAs in animal nutrition. CAB International, Wallingford, UK, pp 411–426

Ihara T, Tsujikawa T, Fujiyama Y, Bamba T (2000) Regulation of PepT1 peptide transporter expression in the rat small intestine under malnourished conditions. Digestion 61:59–67

Janmaat M, Giaccone G (2003) Small-molecule epidermal growth factor receptor tyrosine kinase inhibitors. Oncologist 8:576–586

Jepson M, Bates P, Broadbent P, Pell J, Millward D (1988) Relationship between glutamine concentration and protein synthesis in rat skeletal muscle. Am J Physiol 255:E166–E172

Kajikawa K, Yasui W, Sumiyoshi H, Yoshida K, Nakayama H, Ayhan A et al (1991) Expression of epidermal growth factor in human tissues. Immunohistochemical and biochemical analysis. Virchows Arch A Pathol Anat Histopathol 418:27–32

Kats L, Nelssen J, Tokach M, Goodband R, Hansen J, Laurin J (1994) The effect of spray-dried porcine plasma on growth performance in the early-weaned pig. J Anim Sci 72:2075–2081

Kelly D, McFadyen M, King TP, Morgan PJ (1992) Characterization and autoradiographic localization of the epidermal growth factor receptor in the jejunum of neonatal and weaned pigs. Reprod Fertil Dev 4:183–191

Kennedy DJ, Leibach FH, Ganapathy V, Thwaites DT (2002) Optimal absorptive transport of the dipeptide glycylsarcosine is dependent on functional Na+/H+ exchange activity. Eur J Physiol 445:139–146

Kilberg MS, Haussinger D (1992) AA transport in liver. In: Kilberg MS, Haussinger D (eds) Mammalian AA transport. Plenum Press, New York, pp 133–148

Kim Y, Uotani S, Pierroz D, Flier J, Kahn B (2000) In vivo administration of leptin activates signal transduction directly in insulin-sensitive tissues: overlapping but distinct pathways from insulin. Endocrinology 141:2328–2339

Koeln LL, Schlagheck TG, Webb KE Jr (1993) AA flux across the gastrointestinal tract and liver of calves. J Dairy Sci 76:2275–2285

Krehbiel C, Matthews J (2003) Absorption of AAs and peptides. In: D'Mello JPF (ed) AAs in animal nutrition. CAB International, Wallingford, UK, pp 41–70

Le Bacquer O, Laboisse C, Darmaun D (2002) Glutamine preserves protein synthesis and paracellular permeability in Caco-2 cells submitted to "luminal fasting". Am J Physiol 285:G128–G136

Li C, Buettger C, Kwagh J, Matter A, Daikhin Y, Nissim IB, Collins HW, Yudkoff M, Stanley CA, Matschinsky FM (2004) A signaling role of glutamine in insulin secretion. J Biol Chem 279:13393–13401

Liang R, Fei YJ, Prasad PD, Ramamoorthy S, Han H et al (1995) Human intestinal H+/peptide cotransporter. Cloning, functional expression, and chromosomal localization. J Biol Chem 270:6456–6463

Lima A, Carvalho G, Figueiredo A, Gifoni A, Soares A, Silva E, Guerrant R (2002) Effects of an alanyl-glutamine-based oral rehydration and nutrition therapy solution on electrolyte and water absorption in a rat model of secretory diarrhea induced by cholera toxin. Nutrition 18:458–462

Lindemann M, Cromwell G, Monegue H, Cook H, Soltwedel K, Thomas S, Easter R (2000) Feeding value of an enzymatically digested protein for early-weaned pigs. J Anim Sci 78:318–327

MacDonald RS, Thornton WH Jr, Bean TL (1993) Insulin and IGF-1 receptors in a human intestinal adenocarcinoma cell line (Caco-2): regulation of Na+ glucose transport across the brush border. J Recept Res 13:1093–1113

Marks SL (1998) The principles and practical application of enteral nutrition. Vet Clin N Am Small Anim Pract 28:677–708

Marrs TC, Addison JM, Burston D, Matthews DM (1975) Changes in plasma AA concentrations in man after ingestion of an AA mixture simulating casein, and a tryptic hydrolysate of casein. Br J Nutr 34:259–265

Maruyama S, Nonaka I, Tanaka H (1993) Inhibitory effects of enzymatic hydrolysates of collagen and collagen-related synthetic peptides on fibrinogen/thrombin clotting. Int J Biochem Biophys 1164:215–218

Matsufuji H, Matsui T, Seki E, Osajima K, Nakashima M, Osajima Y (1994) Angiotensin I-converting enzyme inhibitory peptides in an alkaline protease hydrolyzate derived from sardine muscle. Biosci Biotechnol Biochem 58:2244–2245

Matsui T, Li CH, Tanaka T, Maki T, Osajima Y, Matsumoto K (2000) Depressor effect of wheat germ hydrolysate and its novem angiotensin I-converting enzyme inhibitory peptide, Ile-Val-Tyr, and the metabolism in rat and human plasma. Biol Pharm Bull 23:427–431

Matthews DM (1991) Protein absorption: development and present state of the subject. Wiley-Liss, New York

Matthews JC (2000a) AA and peptide transport systems. In: D'Mello JPF (ed) Farm animal metabolism and nutrition. CAB International, Wallingford, UK, pp 3–22

Matthews JC (2000b) Peptide absorption: where peptides fit in protein nutrition and metabolism. In: Lyons TP, Jacques KA (eds) Biotechnology in the feed industry. Proceeding's of Alltech's sixteenth annual symposium. Nottingham University Press, Nottingham, UK, pp 357–368

Matthews DM, Addison JM, Burston D (1974) Evidence for active transport of the dipeptide carnosine (β-alanyl-L-histidine) by hamster jejunum in vitro. Clin Sci Mol Med 46:693–705

McCormick ME, Webb KE Jr (1982) Plasma free, erythrocyte free and plasma peptide AA exchange of calves in steady state and fasting metabolism. J Nutr 112:276–282

Meridith D, Boyd CA (2000) Structure and function of eukaryotic peptide transporters. Cell Mol Life Sci 57:754–778

Miyoshi S, Ishikawa H, Kaneko T, Fukui F, Tanaka H, Maruyama S (1991) Structures and activity of angiotensin-converting enzyme inhibitors in an alpha-zein hydrolysate. Agric Biol Chem 55:1313–1318

Monchi M, Rerat AA (1993) Comparison of net protein utilization of milk protein milk enzymatic hydrolysates and free AA mixtures with a close pattern in the rat. JPEN J Parenter Enteral Nutr 17:355–363

Moughan PJ, Pedraza M, Smith WC, Williams M, Wilson MN (1990) An evaluation with piglets of bovine milk, hydrolyzed bovine milk, and isolated soybean proteins included in infant milk formulas. I. effect on organ development, digestive enzyme activities, and AA digestibility. J Pediatr Gastrointestinal Nutr 10:385–394

Muller U, Brandsch M, Prasad P, Fei Y, Ganapathy V, Leibach F (1996) Inhibition of the H+/peptide cotransporter in the human intestinal cell line Caco-2 by cyclic AMP. Biochem Biophys Res Commun 218:461–465

Nagai K, Suds T, Kawasaki K, Mathuura S (1986) Action of carnosine and β-alanine on wound healing. Surgery 100:815–821

Nakamura Y, Yamamoto N, Sakai K, Takano T (1995) Antihypertensive effect of sour milk and peptides isolated from it that are inhibitors of angiotensin I-converting enzyme. J Dairy Sci 78:1253–1257

Nakano T, Simatani M, Murakami T, Sato N, Idota T (1994a) Digestion and absorption of enzymatically hydrolyzed whey protein. J Jpn Soc Nutr Food Sci 47:195–201

Nakano T, Simatani M, Murakami T, Sato N, Idota T (1994b) Utilization of nitrogen in enzymatically hydrolyzed whey protein. J Jpn Soc Nutr Food Sci 47:203–208

Naruhashi K, Sai Y, Tamai I, Susuki N, Tsuji A (2002) PepT1 mRNA expression is induced by starvation and its level correlated with absorptive transport of cefadroxil longitudinally in the rat intestine. Pharm Res 19:1417–1423

Nielsen C, Brodin B (2003) Di/tri-peptide transporters as drug delivery targets: regulation of transport under physiological and patho-physiological conditions. Curr Drug Targets 4:373–388

Nielsen CU, Amstrup J, Steffansen B, Frokjaer S, Brodin B (2001) Epidermal growth factor inhibits glycylsarcosine transport and hPepT1 expression in a human intestinal cell line. Am J Physiol Gastrointest Liver Physiol 281:G191–G199

Nielsen CU, Amstrup J, Nielsen R, Steffansen B, Frokjaer S, Brodin B (2003) Epidermal growth factor and insulin short-term increase hPepT1-mediated glycylsarcosine uptake in Caco-2 cells. Acta Physiol Scand 178:139–148

Nunez M, Bueno J, Ayudarte M, Almendros A, Rios A, Suarez M, Gil A (1996) Dietary restriction induces biochemical and morphometric changes in the small intestine of nursing piglets. J Nutr 126:933–944

Ogihara H, Suzuki T, Nagamachi Y, Inui KI, Takata K (1999) Peptide transporter in the rat small intestine: ultrastructural localization and the effect of starvation and administration of AAs. Histochem J 31:169–174

Palacin M, Estevez R, Bertran J, Zorzano A (1998) Molecular biology of mammalian plasma membrane AA transporters. Physiol Rev 78:969–1054

Pan Y, Wong EA, Bloomquist JR, Webb KE Jr (2001) Expression of a cloned ovine gastrointestinal peptide transporter (oPepT1) in *Xenopus* oocytes induces uptake of oligopeptides in vitro. J Nutr 131:1264–1270

Pettigrew JE, Harmon BG, Simon J, Baker DH (1977) Milk proteins for artificially reared piglets: II. Comparison to a skim milk hydrolysate. J Anim Sci 44:383–388

Playford RJ, Hanby AM, Gschmeissner S, Pfeiffer LP, Wright NA, McGarrity T (1996) The epidermal growth factor (EGF-R) is present on the basolateral, but not apical, surface of enterocytes in the human gastrointestinal tract. Gut 39:262–266

Pluske J, Thompson M, Atwood C, Bird P, Williams I, Hartmann P (1996) Maintenance of villus height and crypt depth, and enhancement of disaccharide digestion and monosaccharide absorption, in piglets fed on cows' whole milk after weaning. Br J Nutr 76:409–422

Poullain MG, Cezard JP, Marche C, Roger L, Mendy F, Broyart JP (1989a) Dietary whey proteins and their peptides or AAs: effects on the jejunal mucosa of starved rats. Am J Clin Nutr 49:71–76

Poullain MG, Cezard JP, Roger L, Mendy F (1989b) Effect of whey proteins, their oligopeptide hydrolysates and free AA mixtures on growth and nitrogen retention in fed and starved rats. JPEN J Parenter Enteral Nutr 13:382–386

Reeds PJ, Burrin DG, Stoll B, Jahoor F, Wykes L, Henry J, Frazer ME (1997) Enteral glutamate is the preferential source for mucosal glutathione synthesis in fed piglets. Am J Physiol 237:E408–E415

Remillard R, Armstrong P, Davenport D (2000) Assisted feeding in hospitalized patients: enteral and parenteral nutrition. In: Hand MS, Thatcher CD, Remillard RL et al (eds) Small animal clinical nutrition (Ch. 12), 4th edn. Walsworth, Marceline, MO

Rerat A (1993) Nutritional supply of proteins and absorption of their hydrolysis products: consequences on metabolism. Proc Nutr Soc 52:335–344

Rerat A (1995) Nutritional value of protein hydrolysis products (oligopeptides and FAAs) as a consequence of absorption and metabolism kinetics. Arch Anim Nutr 48:23–36

Rerat A, Nunes CS (1989) AA absorption and production of pancreatic hormones in non-anaesthetized pigs after duodenal infusions of a milk enzymatic hydrolysate or of FAAs. J Nutr 60:121–136

Rerat A, Nunes CS, Mendy F, Vaissade P, Vaugelade P (1992) Splanchnic fluxes of AAs after duodenal infusion of carbohydrate solutions containing free-AAs or oligopeptides in the non-anaesthetized pig. Br J Nutr 68:111–138

Ririe DG, Roberts PR, Shouse MN, Zaloga GP (2000) Vasodilatory actions of the dietary peptide carnosine. Nutrition 16:168–172

Roberts P, Black K, Santamauro J, Zaloga G (1998) Dietary peptides improve wound healing following surgery. Nutrition 14:266–269

Rocha F, Musch M, Lishanskiy L, Bookstein C, Sugi K, Xie U, Chang E (2003) IFN-γ downregulates expression of Na+/H+ exchangers NHE2 and NHE3 in rat intestine and human Caco-2/bbe cells. Am J Physiol Cell Physiol 280:C1224–C1232

Rubio-Aliaga I, Daniel H (2002) Mammalian peptide transporters as targets for drug delivery. Trends Pharmacol Sci 23:434–440

Saiga A, Okumura T, Makihara T, Katsuta S, Shimizu T, Yamada R, Nishimura T (2003) Angiotensin I-converting enzyme inhibitory peptides in hydrolyzed chicken breast muscle extract. J Agric Food Chem 51:1741–1745

Saito H, Okuda M, Terada T, Sasaki S, Inui K (1995) Cloning and characterization of a rat H+/peptide cotransporter mediating absorption of beta-lactam antibiotics in the intestine and kidney. J Pharmacol Exp Ther 275:1631–1637

Samonina G, Ashmarin I, Lyapina L (2002) Glyproline peptide family: review on bioactivity and possible origins. Pathophysiology 8:229–234

Sarosiek J, Bilski J, Murty V, Slomiany A, Slomiany BL (1988) Role of salivary epidermal growth factor in the maintenance of physiochemical characteristics of oral and gastric mucosal mucus coat. Biochem Biophys Res Commun 152:1421–1427

Sarosiek J, Feng T, McCallum RW (1991) The interrelationship between salivary epidermal growth factor and the functional integrity of the esophageal mucosal barrier in the rat. Am J Med Sci 302:359–363

Satoh J, Tsujikawa T, Fujiyama Y, Bamba T (2003a) Nutritional benefits of enteral alanyl-glutamine supplementation on rat small intestine damage induced by cyclophosphamide. J Gastroenterol Hepatol 18:719–725

Satoh J, Tsujikawa T, Fujiyama Y, Bamba T (2003b) Eneral alanyl-glutamine supplement promotes intesinal adaptation in rats. Int J Mol Med 12:615–620

Schweiger M, Steffl M, Amselgruber W (2003) Differential expression of EGF receptor in the pig duodenum during the transition phase from maternal milk to solid food. J Gastroenterol 38:636–642

Seal CJ, Parker DS (1991) Isolation and characteriation of circulating low molecular weight peptides in steer, sheep and rat portal and peripheral blood. Comp Biochem Physiol 99B:679–685

Sekikawa S, Kawai Y, Fujiwara A, Takeda K, Tegoshi T, Uchikawa R, Yamada M, Arizono N (2003) Alterations in hexose, AA and peptide transporter expression in intestinal epithelial cells during Nippostrongylus brasiliensis infection in the rat. Int J Parasitol 33:1419–1426

Silk DBA, Perrett D, Clark ML (1973a) Intestinal transport of two dipeptides containing the same two neutral AAs in man. Clin Sci Mol Med 45:291–299

Silk DBA, Marrs TC, Addison JM, Burston D, Clark ML, Matthews DM (1973b) Absorption of AAs from an AA mixture simulating casein and a tryptic hydrolysate of casein in man. Clin Sci Mol Med 45:715–719

Silk DBA, Clark ML, Marrs TC, Addison JM, Burston D, Matthews DM (1975) Jejunal absorption of an AA mixture simulating casein and an enzyme hydrolysate of casein prepared for oral administration to normal adults. Br J Nutr 33:95–100

Silk DBA, Chung YC, Berger KL, Conley K, Beigler M, Sleisenger MH, Spiller GA, Kim YS (1979) Comparison of oral feeding of peptide and AA meals to normal human subjects. Gut 20:291–299

Silk DBA, Fairclough P, Clark M, Hagerty J, Marrs T, Addison J, Burston D, Clegg K, Matthews D (1980) Use of a peptide rather than free AA nitrogen source in chemically defined "elemental" diets. JPEN J Parenter Enteral Nutr 4:548–553

Sleisenger MH, Pelling D, Burston D, Matthews DM (1977) AA concentrations in portal venous plasma during absorption from the small intestine of the guinea pig of an AA mixture simulating casein and a partial enzymic hydrolysate of casein. Clin Sci Mol Med 52:259–267

Sobhani I, Bado A, Vissuzaine C, Buyse M, Kermorgant S, Laigneau JP, Attoub S, Lehy T, Henin D, Mignon M, Lewin MJ (2000) Leptin secretion and leptin receptor in the human stomach. Gut 47:178–183

Stoll B, Henry J, Reeds P, Yu H, Yahoor F, Burrin D (1998) Catabolism dominates the first-pass intestinal metabolism of dietary essential AAs in milk protein-fed piglets. J Nutr 128:606–614
Sun B, Zhao X, Wang G, Li N, Li J (2003a) Hormonal regulation of dipeptide transporter (PepT1) in Caco-2 cells with normal and anoxial/reoxygenation management. World J Gastroenterol 9:808–812
Sun B, Zhao X, Wang G, Li N, Li J (2003b) Changes of biological functions of dipeptide transporter (PepT1) and hormonal regulation in severe scald rats. World J Gastroenterol 12:2782–2785
Tagari H, Webb K Jr, Theurer B, Huber T, DeYoung D, Cuneo P et al (2004) Portal drained viceral flux, hepatic metabolism, and mammary uptake of free and peptide-bound AAs and milk AA output in dairy cows fed diets containing corn grain steam flaked at 360 or steam rolled at 490 g/L. J Dairy Sci 87:413–430
Tanaka H, Miyamoto KI, Morita K, Haga H, Segawa H, Shiraga T, Fujioka A, Kouda T, Taketani Y, Hisano S, Fukui Y, Kitagawa K, Takeda E (1998) Regulation of the PepT1 peptide transporter in the rat small intestine in response to 5-fluorouracil-induced injury. Gastroenterology 114:714–723
Thamotharan M, Bawani S, Zhou X, Adibi S (1999a) Functional and molecular expression of intestinal oligopeptide transporter (Pept-1) after a brief fast. Metabolism 48:681–684
Thamotharan M, Bawani S, Zhou X, Adibi S (1999b) Hormonal regulation of oligopeptide transporter pept-1 in a human intestinal cell line. Am J Physiol 276:C821–C826
Torres N, Lopez G, De Santiago S, Hutson S, Tovar A (1998) Dietary protein level regulates expression of the mitochondrial branched-chain aminotransferase in rats. J Nutr 128:1368–1375
Tsuji A, Tamai I (1996) Carrier-mediated intestinal transport of drugs. Pharm Res 13:963–977
Vaughan TJ, Pascall JC, James PS, Brown KD (1991) Expression of epidermal growth factor and its mRNA in pig kidney, pancreas, and other tissues. Biochem J 279:315–318
Watanabe K, Terada K, Jinriki T, Sato J (2003) Effect of insulin on cephalexin uptake and transepithelial transport in the human intestinal cell line Caco-2. Eur J Pharm Sci 21:87–93
Wollheim CB, Biden TJ (1986) Signal transduction in insulin secretion: comparison between fuel stimuli and receptor agonists. Ann NY Acad Sci 488:317–333
Woods CA, Matthews A, Etienne N, Davenport G, Matthews JC (2001) Molecular identification and biochemical characterization of canine PepT1 function in MDCK cells. FASEB J 15(5):A829
Wu G (1998) Intestinal AA catabolism. J Nutr 128:1249–1252
Wu G, Meier S, Knabe D (1996) Dietary glutamine supplementation prevents jejunal atrophy in weaned pigs. J Nutr 126:2578–2584
Yamamoto N (1997) Antihypertensive peptides derived from food proteins. Biopolymers 43:129–143
Yamamoto S, Korin T, Anzai M, Wang MF, Hosoi A, Abe A, Kishi K, Inoue G (1985) Comparative effects of protein, protein hydrolysate and AA diets on nitrogen metabolism of normal, protein-deficient, gastrectomized or hepatectomized rats. J Nutr 115:1436–1446
Yokoyama K, Chiba H, Yoshikawa M (1992) Peptide inhibitors for angiotensin I-converting enzyme from thermolysin digest of dried bonito. Biosci Biotechnol Biochem 56:1541–1545
Zaloga GP, Ward KA, Prielipp RC (1991) Effect of enteral diets on whole body and gut growth in unstressed rats. JPEN J Parenter Enteral Nutr 15:42–47
Zanghi BM, Sipe G, Davenport G, Matthews JC (2004) Evaluation of glycylsarcosine and cefadroxil as substrates for non-invasive determination of canine small intestine PepT1 capacity and demonstration that maximal cefadroxil absorption occurs when consumed 4 h after meal ingestion [Abstract]. J Anim Sci 82(Suppl 1):245
Zhao XT, McCamish MA, Miller RH, Wang L, Lin HC (1997) Intestinal transit and absorption of soy protein in dogs depend on load and degree of protein hydrolysis. Am Soc Nutr Sci 127:2350–2356
Ziegler F, Ollivier JM, Cynober L, Masini JP, Coudray-Lucas C, Levy E, Giboudeau J (1990) Efficiency of enteral nitrogen support in surgical patients: small peptides υ non-degraded proteins. Gut 31:1277–1283

Ziegler F, Nitenberg G, Coudray-Lucas C, Lasser P, Giboudeau J, Cynober L (1998) Pharmacokinetic assessment of an oligopeptide-based enteral formula in abdominal surgery patients. Am J Clin Nutr 67:124–128

Ziegler F, Evans M, Fernandez-Estivariz C, Jones DP (2003) Trophic and cytoprotective nutrition for intestinal adaptation, mucosal repair, and barrier function. Annu Rev Nutr 23:229–261

Ziemlanski S, Cieslakowa D, Kunachowicz H, Palaszewska M (1978) Balanced Intraintestinal Nutrition: digestion, absorption and biological value of selected preparations of milk proteins. Acta Physiol Pol 29:543–556

Zijlstra RT, Mies AM, McCracken BA, Odle J, Gaskins HR, Lien EL, Donovan SM (1996) Short-term metabolic reponses do not differ between neonatal piglets fed formulas containing hydrolyzed or intact soy proteins. J Nutr 126:913–923

Chapter 10
Protein Hydrolysates/Peptides in Animal Nutrition

Jeff McCalla, Terry Waugh, and Eric Lohry

Abstract The use of protein hydrolysates as an important nutrient for growth and maintenance has been increasing in animal nutrition. Although animal proteins and protein hydrolysates are widely used however, recently vegetable protein hydrolysates are gaining importance. This chapter reviews the use of protein hydrolysates developed by enzyme hydrolysis and by solid state fermentation process in animal nutrition especially for piglets and compares it with the standard products such as plasma and fishmeal.

Keywords Peptides • Protein hydrolysates • Amino acids • Plasma • Fish meal • Absorption • Nutrition • Livestock • Piglets • Villi • Growth • Pepsoygen • Carry over effect

Introduction

The goal of any modern livestock production facility is to enhance profitability by capturing biological efficiencies and utilizing good management practices. One of the single largest costs involved in producing food animals is the cost of feed. The range of feed costs as a percentage of total cost differs from species to species, but it is well known that if feed costs can be positively impacted by improved feeding efficiencies, the bottom line is likely to show a benefit.

Protein, in particular, plays a critical role in the growth and development of young animals. It is in these early stages of life, providing easily digestible sources of protein can set the stage for improving long term growth performance. The use of hydrolyzed proteins can provide benefits on performance necessary for livestock

J. McCalla, T. Waugh, and E. Lohry (✉)
Nutra-Flo Protein and Biotech Products, Sioux City, IA 51106, USA
e-mail: cejlohry@nfprotein.com

V.K. Pasupuleti and A.L. Demain (eds.), *Protein Hydrolysates in Biotechnology*,
DOI 10.1007/978-1-4020-6674-0_10,

producers to gain an economic advantage. This chapter will give more insight into the practical applications and the role of hydrolyzed proteins (peptides) in animal diets.

Protein Hydrolysates/Peptides in Livestock Production

Using protein hydrolysates/peptides (short chains of amino acids) in animal diets is a relatively new concept; however the principles behind how they work are well established. It has long been known that proteins must be further broken down into amino acids before an animal can utilize them for normal metabolic functions. What has been largely ignored is the fact that peptides can also be used in an animal diet similar to the way in which free amino acids are used.

Animals absorb smaller peptides the same way in which it absorbs free amino acids, but the route by which peptides reach the target cells is accomplished in a different manner (Fig. 1). The main difference is the fact that specific transport mechanisms are used to accomplish this task. These specific transport mechanisms that are used to transport small peptides are distinctly different from the mechanisms used to transport free amino acids in an animal (Mellor 2000) and this has

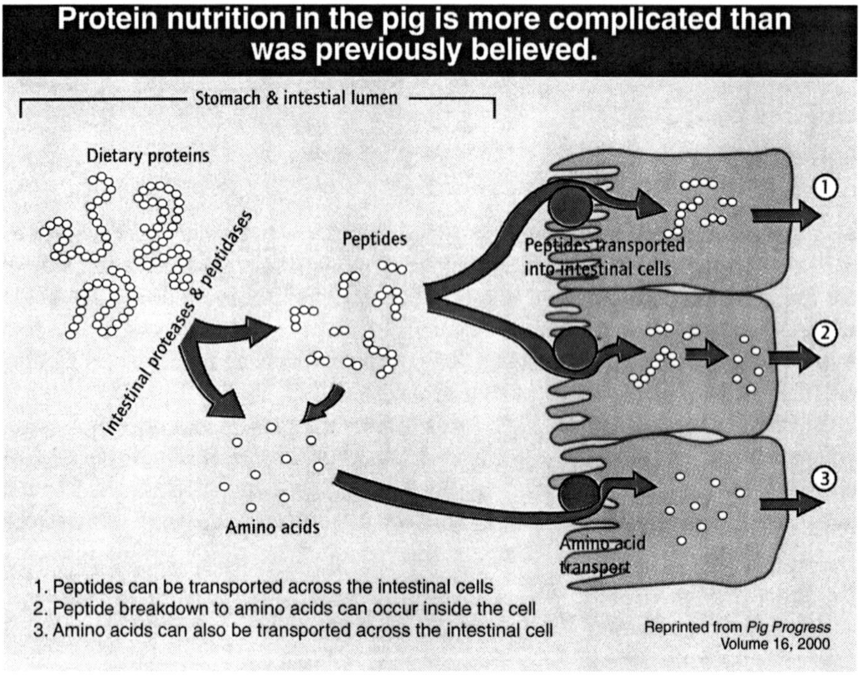

Fig. 1 Protein nutrition in the pig is more complicated than was previously believed

been studied in great detail (Guidotti and Gazzola 1992; Matthews et al. 1996). The degree of specificity that exists for these transport mechanisms has also been evaluated. For instance, certain studies (Silk et al. 1985) have shown that peptides with five or fewer amino acid residues are absorbed with greater efficiency than larger peptides that fall outside of this size range.

Other researchers have suggested that certain amino acids that are often difficult to provide in their free form have much greater availability when provided in the form of a dipeptide (Furst and Stehle 1993). It has also been shown (Rerat et al. 1988) that the rate of absorption of small peptides is greater when compared to the absorption rate for an equivalent amount of free amino acids. This is an important factor to consider when formulating animal diets, since increasing the efficiency and rate of absorption of amino acids can make a critical difference to an animal that may be undergoing some sort of challenge due to its environment or otherwise. Younger animals with digestive disorders, for instance, may benefit from the increased rate of absorption of a peptide when compared to that of free amino acids.

Some studies (Ji et al. 1999) have shown significant increases in the number and size of villi present in the small intestine when pigs were fed peptides compared with other intact proteins commonly used in piglet diets. An increase in the number and size of villi present increases the amount of surface area available for nutrient absorption, which ultimately can improve the efficiency of growth performance. It was determined from this study that the observed improvements in gut health were highly correlated with improvements in growth performance.

Other studies have evaluated the use of hydrolysates that originate from specific raw material sources in an effort to provide a specific function in the diet. An example of such work was demonstrated in a study that evaluated a hydrolysate containing a high level of glutamine for the purposes of enhancing intestinal morphology (Lai et al. 2004). Other researchers have evaluated the effects that specific peptides may have on regulating gastrointestinal function and intake (Froetschel 1996).

A review conducted by Webb et al. (1992), which focused on peptide absorption, indicated that significant differences might exist between various species of animals. Most notably, the lack of understanding of peptide absorption in ruminant species was a major theme of the review. The summary of findings indicates that peptides may have greater quantitative importance in ruminant species than do free amino acids.

As more knowledge is acquired, all of the above factors can play a role in helping nutritionists formulate diets for animals with more precision. Greater precision may enable livestock producers to feed their animals more efficiently and reduce nutrient outputs, which has several positive implications for the environment.

Peptides in Human Nutrition

Peptides have been used in human nutrition for a variety of purposes and applications. Perhaps the most common or well-known use of peptides is in infant nutrition for the purpose of repairing damaged intestinal cell wall structures. A recent review evaluated the use of colostrum and milk-derived peptide growth factors as a treatment for gastrointestinal disorders (Playford et al. 2000). It suggested that peptide growth factors might provide a variety of treatment options for a number of gastrointestinal conditions. It has also been thought that colostrum-derived supplements may serve a useful role in gut adaptation for young children who have undergone intestinal resection.

Colostrum contains a variety of peptide components including: hormones, cytokines, and growth factors. All of these peptide components play different roles in the body. Some peptides have been known to play a beneficial role to those who have compromised immune systems, such as people with HIV infection. The use of colostrum-derived peptides in these situations can be helpful in preventing the onset of infection. In other situations, peptides can be a valuable asset in stimulating the repair process during infection (Sarker et al. 1998).

Peptides from milk have also been used as potential treatment options for a variety of other purposes. Seppo et al. (2003) successfully demonstrated the use of bioactive peptides derived from fermented milk in lowering blood pressure in hypertensive subjects.

Other uses of peptides in human nutrition is their inclusion in hypoallergenic diets. Intact proteins of various sources can cause allergies; however, when that same protein has been hydrolyzed, the allergic reaction can be significantly reduced and possibly eliminated. The same practice also exists in animal diets, especially in companion animals that have grain-related allergies. Using peptides to treat allergies may just be one of the first steps towards finding new uses for hydrolyzed proteins and bioactive peptides in animals. Improving the health of humans and animals by creating products that serve a specific function will likely play a larger role in the future.

Human Nutrition Versus Animal Nutrition

Although nutritional principles are similar for humans and animals, their application is much different. The primary objectives for human nutrition are to improve health and increase life expectancy. The objectives for companion animals would be similar. However, for production animals, nutritional objectives are much different. The nutritional objectives for production animals focus on strategies to optimize food production, e.g., increasing milk production in dairy cows or post weaning weight gain in pigs. Although genetics significantly impacts the performance of food producing animals, environmental effects are a much greater determinant of performance of which nutrition and health are the most important.

Performance parameters influenced by nutrition include:

- In cattle production:
  1. Percent calf crop – calves born live/number cows in the herd
  2. Two hundred and five day calf weight – calf weight at weaning and at 205 days post-calving
  3. Average daily weight gain of feedlot cattle

- In swine production:
  1. Number of piglets weaned/litter
  2. Piglet weaning weight – typically pigs are 21 days of age
  3. Average daily weight gain of grower/finisher pigs
  4. Feed efficiency (grams of weight gain/kilograms of feed consumed)

These parameters measure the production efficiencies of particular livestock enterprises and greatly impact enterprise profitability.

Growth Performance of Animals Fed Peptides from Hydrolysates

The young piglet, like a human infant, has immature digestive and immune systems, relying on maternal milk to provide antibodies to achieve passive immunity for protection from disease and efficient early growth performance. The first diet the pig receives post-weaning attempts to continue this trend by formulating highly palatable, complex diets, including high quality proteins, e.g. dietary peptides.

In piglets, dietary peptides have been shown to elicit better overall growth response than dietary plasma in terms of average daily weight gain and gain/unit of feed consumed (Zimmerman 1996a, b). Hydrolysates have also been shown to improve pig growth performance and gut morphology relative to other commonly used protein sources, e.g. fish meal (Ji 1999; Stein 2002) and spray-dried blood cells (Lindemann 1997). Recently, fermented soybean meal, a source of high quality peptides, has been shown to elicit better overall performance than plasma when included in piglet diets (Kim 2003) and hydrolyzed soy protein isolates have been have been shown to be suitable for veal calf diets (Lalles 1995).

Comparing Enzymatically Hydrolyzed Protein to Animal Plasma

In a study conducted by Zimmerman (1996a) of Iowa State University, porcine hydrolysate (co-product of heparin production for the human pharmaceutical industry) was compared to spray-dried plasma protein, an expensive high-quality protein ingredient to determine its effect on piglet growth performance when included in the diet as a protein source in a 5-week feeding trial. Pigs were fed diets that included either plasma, the hydrolysate or the control diet for the first 2 weeks of the trial, and a common diet with neither plasma nor hydrolysate the last 3 weeks. The data showed that piglets fed porcine hydrolysate had better overall growth performance than piglets fed plasma. A carryover effect was observed with growth performance increasing in the hydrolysate-fed pigs when they were fed the common diet during the weeks 3, 4 and 5.

Enzymatically Hydrolyzed Protein as a Replacement for Fish Meal

A 5-week feeding trial conducted by Stein (2002) at South Dakota State University was conducted to evaluate the efficacy of porcine hydrolysates on the growth performance of piglets weaned at 20 days of age. Piglets fed porcine hydrolysates, weeks 1–3 had significantly better growth rate and feed efficiency (weight gain/unit of feed consumed) than piglets fed high quality fishmeal with equal feed intake. Stein stated that the data suggest that improved growth performance of the hydrolysate-fed piglets was the result of improved gut health or perhaps nutrient uptake. In the trial reviewed in Figs. 2 and 3, improved gut health was demonstrated with improved villus height/crypt depth measurements in piglets fed hydrolysates compared to piglets fed fishmeal.

Enzymatically Hydrolyzed Proteins Versus Spray-Dried Blood Cells

A 4-week feeding trial conducted at the University of Kentucky by Lindemann (1997) evaluated the effects of enzymatic hydrolysates on growth performance of 20 day old piglets. Lindemann's data showed that the cumulative average daily weight gain was greater for piglets fed hydrolysates as compared to piglets fed spray-dried blood cells. Additionally, the hydrolysate-fed piglets had superior feed utilization compared to pigs fed the diet containing spray-dried blood cells. Enzymatic hydrolysates are a

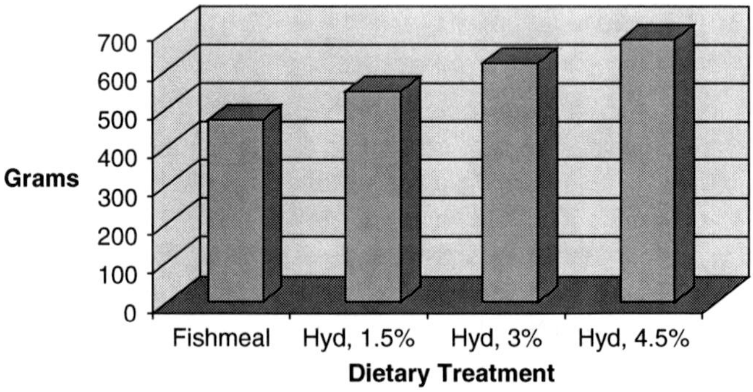

Fig. 2 Porcine hydrolysate effects on average daily gain (Stein 2002)

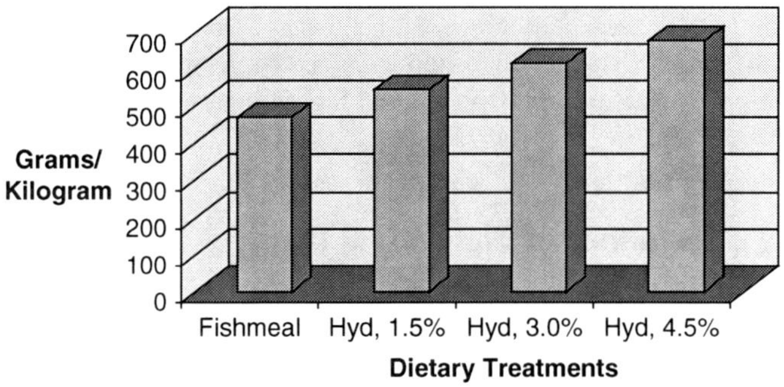

Fig. 3 Porcine hydrolysate effects on feed efficiency (Stein 2002)

suitable replacement for spray-dried blood cells with an advantage in feed utilization (Lindemann 1997).

Dried Whey Compared to Enzymatically Hydrolyzed Animal Proteins

Zimmerman (1996b) compared porcine hydrolysate to dried whey in the diets of 21 day old piglets, in a 5 week-feeding trial. Dried whey is a common ingredient included in weanling pig diets because it furnishes high quality protein and lactose, the carbohydrate of choice for weanling pigs (Zimmerman 1996b). Therefore, in this study, porcine hydrolysate and lactose were included in the diet at the expense of dried whey. The experimental diets were fed for 2 weeks and a common diet was fed for 3 weeks. Piglets fed the hydrolysate had significantly higher average daily gain and feed intake. Again, a carryover effect was observed with increased growth rate in the pigs previously fed porcine hydrolysate when they were on the common diet the final 3 weeks of the trial.

Hydrolyzed Soy Protein Isolate in Veal Calf Diets

A study conducted by Lalles et al. (1995) in France demonstrated that hydrolyzed soy protein isolate is a suitable replacement for high cost skim milk powder as a protein source in calf milk replacers. Hydrolyzed soy protein isolate plus whey or heated soy flour plus whey provided 56% and 72% of the dietary proteins at the expense of skim milk powder in the milk replacer diet of 1 month old calves. Digestive function was evaluated by measurement of ruminal pH, plasma kinetics of triglyceride and glucose concentrations in the soy diets. Post-prandial changes in concentrations of triglycerides and glucose in plasma suggested a lack of abomasal clotting with both diets, which allows for a slower release of nutrients (Petit 1987). However, soy protein isolate supported satisfactory growth and carcass qualities, while the soy flour was poorly digested and highly immunogenic.

Growth Performance of Piglets Fed Peptides from Soy Fermentation

Soybean meal is the most popular protein source in the animal feed industry. Relatively high protein content and wide availability make soybean meal a good source of protein in animal diets. However, use of soybean meal is rather limited to adult animals due to inefficient digestibility of soy proteins by young animals (Li et al. 1990). Fermented dehulled soybean meal is also an excellent source of high quality peptides. Short-chained peptides are produced by selective microorganisms in a highly controlled, solid-state fermentation process. Additionally, this process greatly reduces the levels of anti-nutritional factors in soybean meal, e.g. alpha galactosides and trypsin inhibitor (Kim 2003). Figure 4 shows the distribution of peptides of three different soy products. Using electrophoresis is one way to show how effectively the fermentation process hydrolyzes soy protein, resulting in highly digestible peptides.

Dr. Sung Woo Kim (2003) conducted a series of studies at Texas Tech University to determine the efficacy of fermented soybean meal as a protein source for early-weaned piglets. He compared fermented soybean meal to soybean meal, dried whey and spray-dried plasma protein. The data showed that pigs fed diets with 3% and 6% inclusion of fermented soybean meal at the expense of dried skim milk performed equally as compared to pigs fed dried skim milk in terms of weight gain and gain/unit of feed consumed.

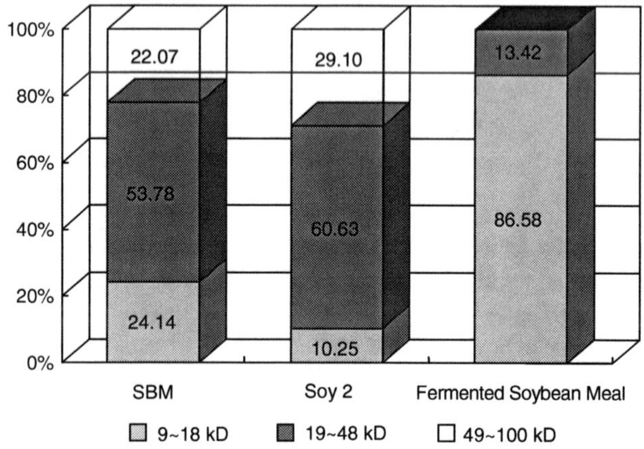

Fig. 4 Peptide composition of three soy products (Kim 2003) (SBM= soybean meal)

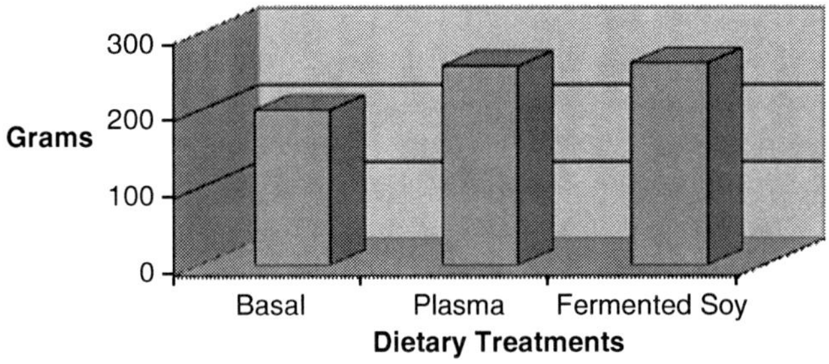

Fig. 5 Fermented soy effects on piglet average daily gain (Kim 2003)

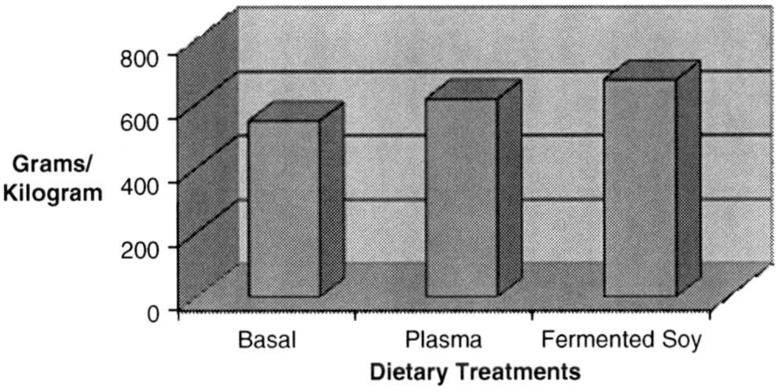

Fig. 6 Fermented soy effects on piglet feed efficiency (Kim 2003)

When fermented soybean meal was compared to spray-dried plasma protein week 1, post-weaning growth performance was better for the plasma-fed pigs (Figs. 5 and 6). However, during weeks 2 and 3 post-weaning growth performance improved with pigs fed fermented soy compared to those fed plasma, suggesting the benefits of feeding fermented soybean meal can be achieved after feeding for at least a 2-week period (Kim 2003)

Carryover Effect

In piglet trials where experimental diets are fed for 2–3 weeks, the piglets are fed many times a common diet for 2 or more additional weeks to determine the duration of the hydrolysates effect, i.e., whether they elicit a positive carryover effect. In Zimmerman's trials comparing porcine hydrolysates to dried whey and plasma, pigs previously fed porcine hydrolysate had superior growth performance during the period where all pigs were on a common diet containing no hydrolysate, whey or plasma (Fig. 7).

Why does a positive carryover effect occur? Improved gut health and perhaps improved nutrient uptake in pigs fed hydrolysates are the most likely reasons for this performance response (Stein 2002). In a study by Poullain et al. (1989) in France, rats were starved for 72 h, followed by re-feeding with whole whey protein, whey protein hydrolysate or an amino acid mixture. The trial indicated that the rats fed with the hydrolysate or amino acid diets had significantly greater villus height and disaccharidase activities than rats fed with the whole whey protein.

Another trial conducted at the Agricultural University of China, Beijing by Ji et al (1999) compared porcine hydrolysate to fishmeal and plasma in a piglet growth and gut morphology assay. Piglets fed the hydrolysate diet had significantly better growth performance and villus height/crypt depth measurements than the fish meal-fed pigs and equal to the plasma-fed pigs. Electron microscopy photographs (Figs. 8 and 9) show the effects of fish meal and porcine hydrolysate on the villi of

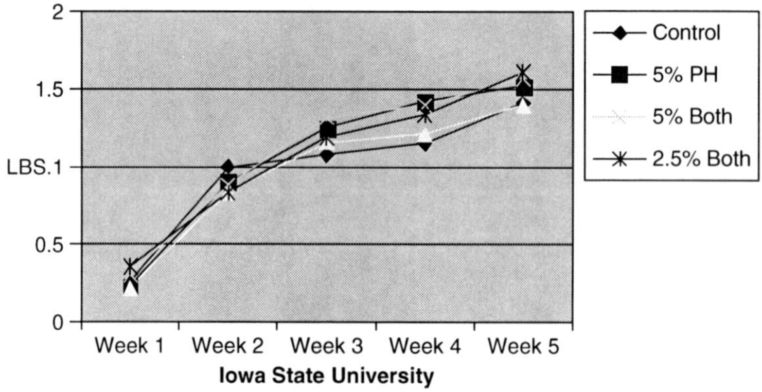

Fig. 7 Porcine Hydrolysate (PH) carryover effect (Zimmerman 1996a)

Fig. 8 Villi from Fishmeal-fed Pigs (Ji et al. 1999)

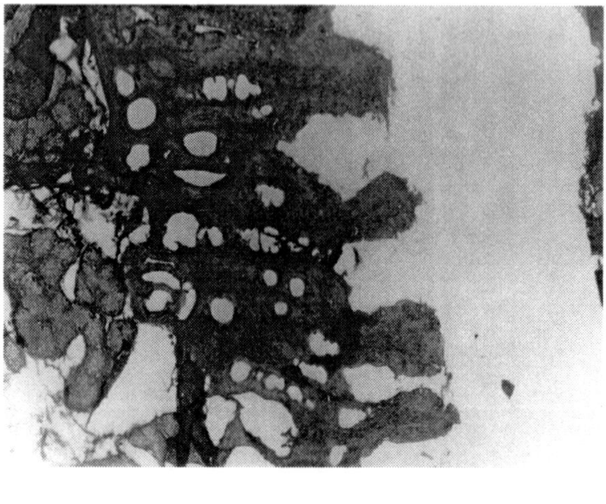

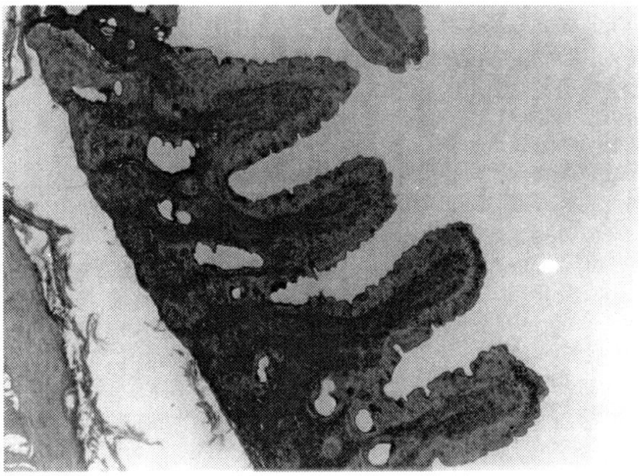

Fig. 9 Villi from Hydrolysate-fed Pigs (Ji et al. 1999)

the gut. The villi of the pigs fed the hydrolysate are elongated and well defined, while the villi from the fishmeal-fed pigs are blunted with tissue being sloughed. The gut morphology of the hydrolysate-fed pigs is much more conducive to efficient nutrient absorption and subsequent growth performance. Gut morphology and growth performance are highly correlated (Tang et al. 1999).

References

Froetschel MA (1996) Bioactive peptides in digesta that regulate gastrointestinal function and intake. J Anim Sci 74:2500–2508

Furst P, Stehle P (1993) The potential use of parenteral dipeptides in clinical nutrition. Nutr Clin Pract 8(3):106–114

Guidotti GG, Gazzola GC (1992) Amino acid transporters: systematic approach and principles of control. In: Kilberg MS, Haussinger D (eds) Mammalian amino acid transport. Plenum, New York, pp 3–29

Ji C (1999) Evaluation of DPS supplement in early weaned pig diets. College of Animal Science, Agricultural University of China, Beijing, PR China, 100094

Ji C et al (1999) Evaluation of DPS supplement in early-weaned pig diets. College of Animal Science, Agricultural University of China, Beijing, PR China, 100094

Kim SW (2003) Novel plant-derived protein for pigs: PepSoyGen. Texas Tech University, Lubbock, TX

Lai CH, Qiao SY, Li D, Piao XS, Bai L, Mao MF (2004) Effects of replacing spray dried Porcine Plasma with Solpro 500 on performance, nutrient digestibility and intestinal morphology of starter pigs. Asian Austral J Anim 17: 2, 237–243

Lalles JP et al (1995 Jan) Hydrolyzed soy protein isolate sustains high nutritional performance in Veal Calves, Institut National de a Recherche Agronomique, Reenes, France. J Dairy Sci 78(1):194–204

Li DF et al (1990) Transient hypersensitivity to soybean meal in the early-weaned pig. J Anim Sci 68(6):1790–1799

Lindemann M (1997) Evaluation of dried porcine solubles in diets for weanling pigs in comparison to spray dried blood cells. University of Kentucky, Lexington, KY

Matthews JC, Pan YL, Wang S, McCollum MQ, Webb KE Jr (1996) Characterization of gastrointestinal amino acid and peptide transport proteins and the utilization of peptides as amino acid substrates by cultured cells (Myogenic and Mammary) and mammary tissue explants. In: Kornegay ET (ed) Nutrient management of food animals to enhance and protect the environment. CRC Press, Boca Raton, FL, pp 55–72

Mellor S (2000) Pig Progress

Owusu-Asiedu A et al (2002) Response of early-weaned pigs to spray-dried porcine or animal plasma-based diets supplemented with egg-yolk antibodies against enterotoxigenic Escherichia coli. J Anim Sci 80:2895–2903

Petit HV, Ivan M, Brisson GJ (1987 Dec) Duodenal flow of digesta in preruminant calves fed clotting or nonclotting milk replacer, Departement de Zootechnie, Universite laval, Quebec. J Dairy Sci 70(12):2570–2576

Playford RJ, Macdonald CE, Johnson WS (2000) Am J Clin Nutr 72(1):5–14

Poullain MG et al (1989) Dietary whey proteins and their peptides or amino acids: effects on the jejunal mucosa of starved rats. Am J Clin Nutr 49:71–76

Rerat A, Simoes Nunes C, Mendy F, Roger L (1988) Br J Nutr 60:121–136

Sarker SA, Casswall TH, Mahalanabis D, Alam NH, Albert MJ, Brussow H, Fuchs GJ, Hammerstrom L (1998) Successful treatment of rotavirus diarrhea in children with immunoglobulin from immunized bovine colostrum. Pediatr Infect Dis J 17:1149–1154

Seppo L, Jauhiainen T, Poussa T, Korpela R (2003) A fermented milk high in bioactive peptides has a blood pressure lowering effect in hypertensive subjects. Am J Clin Nutr 77(2):326–330

Silk DBA, Grimble GK, Rees RG (1985) Proc Nutr Soc 44:63–72

Stein H (2002) The effect of including DPS 50RD and DPS EX in the Phase 2 diets for weanling pigs. South Dakota State University, Brookings, SD

Tang M et al (1999) Effect of segregated early weaning on post-weaning small intestinal development in pigs, Animal Biotechnology Center, Department of Animal & Poultry Science, University of Saskatchewan, Saskatoon, SK, Canada. J Anim Sci 77(12):3191–3200

Webb KE, Matthews JC, DiRenzo DB (1992) Peptide absorption: a review of current concepts and future perspectives. J Anim Sci 70:3248–3257

Zimmerman D (1996) The duration of carry-over growth response to intestinal hydrolysate fed to weanling pigs, Iowa State University, Ames, IA, Experiment 9612

Zimmerman D (1996) Interaction of intestinal hydrolysate and spray-dried plasma fed to weanling pigs, Iowa State University, Ames, IA, Experiment 9615

Chapter 11
Protein Hydrolysates as Hypoallergenic, Flavors and Palatants for Companion Animals

Tilak W. Nagodawithana, Lynn Nelles, and Nayan B. Trivedi

Abstract Early civilizations have relied upon their good sense and experience to develop and improve their food quality. The discovery of soy sauce centuries ago can now be considered one of the earliest protein hydrolysates made by man to improve palatability of foods. Now, it is well known that such savory systems are not just sources for enjoyment but complex semiotic systems that direct the humans to satisfy the body's protein need for their sustenance. Recent developments have resulted in a wide range of cost effective savory flavorings, the best known of which are autolyzed yeast extracts and hydrolyzed vegetable proteins. New technologies have helped researchers to improve the savory characteristics of yeast extracts through the application of Maillard reaction and by generating specific flavor enhancers through the use of enzymes. An interesting parallel exists in the pet food industry, where a similar approach is taken in using animal protein hydrolysates to create palatability enhancers via Maillard reaction scheme. Protein hydrolysates are also utilized extensively as a source of nutrition to the elderly, young children and immuno-compromised patient population. These hydrolysates have an added advantage in having peptides small enough to avoid any chance of an allergenic reaction which sometimes occur with the consumption of larger sized peptides or proteins. Accordingly, protein hydrolysates are required to have an average molecular weight distribution in the range 800–1,500 Da to make them non-allergenic. The technical challenge for scientists involved in food and feed manufacture is to use an appropriate combination of enzymes within the existing economic constraints and other physical factors/limitations, such as heat, pH, and time, to create highly palatable, yet still nutritious and hypoallergenic food formulations.

T.W. Nagodawithana
Esteekay Associates, Inc., Milwaukee, WI 53217, USA

L. Nelles
Kemin Industries, Inc., 2100 Maury Street, Des Moines, IA 50317, USA

N.B. Trivedi (✉)
Trivedi Consulting, Inc., Princeton, NJ 08540, USA
e-mail: nayanbtrivedi@msn.com

V.K. Pasupuleti and A.L. Demain (eds.), *Protein Hydrolysates in Biotechnology*,
DOI 10.1007/978-1-4020-6674-0_11, © Springer Science+Business Media B.V. 2010

Keywords HVP • Hydrolysis of animal protein • Yeast extracts • Chemical and Enzymatic hydrolysis • Food allergies • DH and qualitative analysis

Introduction

Flavor is one of the most important attributes governing the selection of foods we eat. Early civilizations have relied upon their good sense and experience to develop and improve their food quality, which in the process, probably led to the development of their traditional foods. The empirical knowledge they acquired was subsequently transmitted throughout the ages to succeeding civilizations. Today, with this know-how, combined with new technology on the skillful use of ingredients in the art of cooking, fascination with food has become a basic human experience to many. Moreover, it has always been the objective of every chef to develop a well-balanced flavor system to make his or her food preparations a true work of art.

Food is also an important part of our social and physiological well being, in that it is an essential requirement for good health and proper functioning of the human body. It is therefore natural to expect definite links between flavor and nutrition that hold the key to the survival of higher forms of life on this planet. For example, it is common knowledge that certain poisonous compounds are inherently bitter and are eliminated from our food supply by instinct because of their repulsive taste. In contrast, enticing aromas of roasted meat or other savory foods that are essential for good health induce feelings of hunger and stimulate digestive activity. It is, therefore, clear that the desire for a nutritionally important food product could directly be dictated by the correct choice of the flavor delivery system.

Heat treatment plays a very important role in the production of flavor compounds in a wide variety of food products. These heat-induced changes in an aqueous phase are caused by complex thermal reactions between proteins, amino acids, carbohydrates, fats, organic acids, vitamins, etc. The process is referred to as non-enzymatic browning or the Maillard reaction, named after the discoverer. This reaction, that in reality is a complex system of reaction pathways, is important in developing the desirable aroma and non-volatile savory taste chemicals of cooked, roasted, fried or baked foods. Browning and aroma formation accompanying such heat processing are caused primarily by the reaction of carbonyl groups of reducing sugar or related carbonyl compounds with free amino groups of amino acids or peptides. These are the precursors generally present in almost all savory food systems prior to heat processing. The variety of aroma compounds, which can arise from the Maillard reaction generally, varies both in number and complexity based on the reaction conditions. The interested reader may refer to Nagodawithana (1995) for a complete review of the subject.

In most cases, foods require additional flavorings during or after processing to make them more palatable. The significance of a very familiar taste like that of common salt is well known to both food processors and to the consumer. Many other flavorings are now being marketed to improve the flavor characteristics, particularly of convenience foods. Best known ingredients under this category include: yeast extracts, hydrolyzed vegetable proteins (HVPs), soy sauce, cheese powder, cheese flavorings, beef and fish extracts, herbs, spices, etc. Each serves a particular function or several functions in the food system in which they are used. Much of the current research on such base flavors or flavorings is based on the concept that the overall flavor can be improved by balancing the critical flavor components in the food. Additionally, there are certain compounds in our nutritional ecosystem which have little or no flavor of their own, yet have the capability of enhancing the savory flavor already present in the food formulation to give a sense of richness in savor or "meatiness", mouth feel and continuity to the flavor base. These are called flavor enhancers or potentiators.

Among the common savory foods, the best-known fermented savory flavoring that has a long history in the Orient is soy sauce. Although the information relating to this product is fragmentary, the authentic records of the Chinese using soy sauce as a flavoring agent may date back approximately 3,000 years (Prinsen-Geerligs 1896). Today, without doubt, the discovery of soy sauce as rated one of the outstanding achievements made in the area of food science. The popularity of soy sauce as a savory flavoring has clearly soared in the Western world during the last few decades.

It is important to note that there are health benefits associated with the use of protein hydrolysates. For example, the antigenicity of certain proteins are greatly minimized by enzymatic hydrolysis thereby making such hydrolysates suitable for use even in foods for infants who are sensitive to certain food allergies. An interesting parallel exists in the pet food industry where a similar approach is taken in using protein hydrolysates in formulating pet foods, not only to create improved palatability but also to ensure that the products are hypoallergenic.

Modern food processors utilize these flavorings and all the technical resources available at their disposal to skillfully generate the proper balance in the final savory character of their culinary creations. In this chapter, an attempt is made to provide a simplified overview of some of the salient features of the best-known flavor ingredients with a special emphasis on soy sauce, hydrolyzed vegetable proteins, yeast extracts and protein hydrolysates. This subject of savory ingredients has been reviewed earlier by Reed and Nagodawithana (1991), May (1991), and Nagodawithana (1995).

Soy Sauce

Long before recorded history, many communities in the Orient were deprived of animal proteins for a variety of reasons, largely economical, but predominantly, on religious grounds. However, they have been able to enjoy this much-needed savory-meaty character in their diets by use of fermented soy-based products. The best known of these is soy sauce. This product has especially been used as a condiment by such ancient civilizations to improve the palatability of their bland basic diets mainly consisted of rice, fish, bean curd, fermented beans and vegetables. It is now commonly used in food processing as a condiment to enhance color, salt perception, or to improve meat-like savory nuances, particularly by our modern-day culinary chefs.

Although soy sauce originated in China, over time, the empirical knowledge was transmitted to succeeding generations throughout Asia. The per capita consumption of soy sauce is by far the highest in Japan, with China, Taiwan, Malaysia, Korea, Indonesia and the Philippines showing consumption rates in descending order. It must be noted that different civilizations over the years have changed the taste profiles of their own soy sauce to suit their own pallet. Accordingly, Chinese and Japanese soy sauces are substantially different and are rarely substituted one for the other.

Japanese refer to their product as shoyu meaning that it is made by enzymatic digestion of plant proteins in the presence of high salt concentrations (Yokotsuka 1972). The consumption of Japanese shoyu in countries other than Japan has been gradually increasing in the recent past. The United States, which is the largest consumer of Japanese shoyu outside of Japan, has consistently produced well over 10 million liters of genuine Japanese shoyu every year during the last decade.

Production

Soy sauce is a fermented flavoring agent made from soybeans, roasted grains, salt, and water. The organism commonly used for the initial hydrolytic activity of proteins in the substrate are Koji mold, i.e., *Aspergillus oryzae* or *Aspergillus soyae*. This is followed by a vigorous bacterial and yeast fermentation that would produce lactic acid and alcohol.

The fermentation process of soy sauce however, is exceedingly time consuming, requiring several months of fermentation to impart the rich mellow flavor, aroma and the delicate amber color. An option developed lately is the use of an ingredient derived from acid or enzymatic hydrolysis of various proteins.

The fermented material is then pasteurized at a relatively high temperature to generate the characteristic dark reddish-brown color and strong beef-like flavor. Several excellent reviews on soy sauce have appeared in the past (Yokotsuko 1986; Steinkraus 1983; Hesseltine 1983).

Applications

Soy sauce is a flavoring agent because of its high content of peptides and amino acids. These degradation products of proteins together with the new reaction flavors produced during processing enhance the brothy, flavorful character in savory foods. Besides, the high glutamic acid present in soy sauce works synergistically with salt to produce the flavor enhancing effect, thereby offering the "umami" character. Because of such inherent properties, naturally brewed soy sauce heightens the meaty notes in savory dishes while adding depth and color and enhancing aroma to the food without requiring much preparation time. That makes it ideal for use in prepared entrees, soups and sauces. Because of these flavor and flavor enhancing properties, soy sauce has found use in a surprising variety of foods.

Hydrolyzed Vegetable Proteins (HVPs)

In 1886, Julius Maggi, a pioneer in the food industry broke new ground by developing rapid-cooking dehydrated soups (Heer 1991). One of the key ingredients he utilized in his formulation was hydrolyzed plant protein. These hydrolysates yielded the meaty flavoring necessary to make these rapid-cooking soups, meaty and enticing. This eventually evolved into an important business segment in many parts of the world.

HVPs are generally described as flavor donors, flavor enhancers or flavor donor/enhancer combination products. As a flavor donor, they contribute a taste which becomes a distinct feature of the final food base. The high glutamate concentration makes it a flavor enhancer, enriching the naturally occurring flavor of the savory-based foods. These hydrolysates are known for their versatility and ease of application in various food systems. For example, in addition to intensifying naturally occurring savory flavors, they are known to round off and balance the overall savory character in foods. They contain strong flavor intensities offering low usage levels and thus relatively low costs in formulating a wide variety of food products.

Numerous variations of HVPs exist, but one way to categorize them is by color. A light version is likely to be used in poultry, pork and vegetable products while its dark counterpart can be applied in broths, sauces, gravies, meats and stews. A distinct advantage of HVPs is its stability under varying process conditions. Hence, it can be used in canning and freezing processes without any breakdown or change in flavor. It has also been used successfully in baking flavored snacks and crackers at around 180°C.

Recent years have witnessed some decline in the HVPs usage due to reports that it is carcinogenic. Admittedly, some of the trace chemical compounds found previously in HVPs have shown mutagenic properties. However, new technologies have been able to bring the levels of these compounds below 5 ppb in those HVP products that are currently being marketed. Nevertheless, rumors and stories persist for long periods and many product designers seek to avoid or drastically reduce the level of HVPs usage in product formulations.

Production

The most common sources of proteins for HVPs production are de-fatted soy flour, wheat gluten, corn gluten and cottonseeds. These ingredients are selected based on the desired end product and economics. The hydrolysis can be carried out by acid, alkali or by enzymatic treatment. One significant drawback in the enzymatic process is that the hydrolysis is incomplete with the probable generation of bitter peptides. Likewise, the lesser known alkaline hydrolysis often yields an unbalanced amino acid profile, partly due to the destruction of cysteine and arginine, which presents an objectionable sensation to the palate. This subject on enzymatic hydrolysis of proteins will be dealt in more detail, later in this chapter.

The preferred method is acid hydrolysis, which is rapid, low cost, and yields a product with a highly acceptable savory profile. The manufacturing process starts by mixing the appropriate grain proteins in the desired ratio and subsequently treating with a strong acid such as hydrochloric acid. A typical process lasts for 10–12 h at 212°F or 1.5 h under pressure at 250°F. It is an exothermic reaction at elevated temperature and pressure.

The taste of HVPs is rigidly controlled by proper selection and blending of the protein source and appropriate selection of hydrolytic conditions. A high proportion of wheat gluten is preferred because of its high concentration of flavor-enhancing glutamic acid (nearly 38%) contributed by such hydrolysates. During the protein hydrolysis, the peptide bonds are broken leaving free amino acids. Due to the presence of carbonyl compounds, there will be concurrent reactions between the amino and carbonyl groups at high temperature to yield an array of savory flavor compounds.

After the reaction has gone to completion, the mixture is cooled and neutralized with caustic soda or a mixture of soda ash and caustic soda. The salt formed by the acid neutralization contributes to the taste of the product. The liquid medium at this stage is intensely dark in color, primarily due to the formation of sludge, which is sometimes referred to as humin. The product is them filtered, and then bleached and debittered using activated carbon, concentrated and dried. HVPs pastes are less prevalent in the food industry. It is common to add certain flavor enhancing 5′-nucleotides, MSG or caramel color to improve acceptability of the products.

Flavor Characteristic Applications

The purpose of the use of HVPs in food products is primarily for flavor. Its usage level is generally around 1–2% at which point the salt content becomes a self-limiting factor with respect to use level. It provides a significant saving over beef extracts while maintaining the same savory profile. Bouillon cubes, which contain little, if any, beef or chicken extracts, are an obvious example in this regard.

HVPs serve three important functions when incorporated in savory products such as meat: they contribute specific compounds that can fill missing flavor gaps in the food base. They serve as a precursor to develop other desirable flavors during processing and also serve as a flavor enhancer thus intensifying the already existing savory notes in the product.

HVPs can be used as flavor enhancers in such processed foods as soups, sauces, stews, chilis, gravies and in some meat products such as sausages and hotdogs.

Considering the time consuming process for soy sauce production, an option has now developed to partially utilize HVPs in the formulation to make the process cost effective. Such HVPs-based soy sauces bear the words "short term brewed" or "non-brewed" on their labels.

Already, the FDA requires more than just 'hydrolyzed vegetable proteins' on the ingredient label. Regulations require food labels to identify specific protein source, for example 'hydrolyzed corn protein' to clarify its source. The common or the usual name of the food must adequately describe its base nature, characterizing properties or ingredients. General terms such as 'animal' or 'vegetable' are considered unsuitable because protein hydrolysates from different sources are known to serve dissimilar functions. In addition, consumers with special dietary requirements need to know the sources of the additives.

Yeast Extracts

Yeast extracts are the soluble portion of the cell content of yeast following a controlled autolysis or hydrolysis. Best-known extracts are derived from specially selected strains of primary grown baker's yeast. However, similar yeast extracts can also be made from other species of yeast such as torula yeast (*Candida utilis*) grown on ethanol, *Kluyveromyces marxianus* grown on whey, or from spent brewer's yeast. Yeast extracts have become popular during the last few decades in part due to their usefulness as a natural flavoring agent. However, there are now strong trends that indicate an increased demand for second-generation products such as the standard type of flavor enhancers and extracts with exceedingly high flavor enhancing capability.

Yeast extracts are a perfectly natural source for savory flavor for use in a wide variety of food formulations. Its savory character is due to the presence of a variety of amino acids, peptides and reaction flavors generated during autolysis (or hydrolysis) and downstream processing. The choice of the starting material for the production of yeast extracts is directly dictated by the cost and availability of the yeast and the quality expected in the final product.The yeast extracts described earlier are generally considered more as flavoring agents rather than enhancers because their final extracts lack the flavor enhancing 5′-nucleotides. In a typical autolyis process, nucleic acids are converted to 3′-nucleotides because yeast does not have the necessary enzymes to convert RNA to flavor enhancing 5′-nucleotides (Nagodawithana 1994).

Although yeast is used as a source of RNA for making yeast extract containing the flavor enhancing 5′-nucleotides, the RNA content of yeast is highly strain dependent. Among different commercial stains of yeast, *Candida utilis* (torula yeast) has been found to contain the highest level of RNA (10–12%) on a dry solids basis as compared to 5–8% RNA in *Saccharomyces cerevisiae* (baker's yeast). It is however crucial to use special enzymes to breakdown the RNA to release the flavor enhancing 5′-nucleotides. The better known of these flavor potentiators are 5′-guanosine monophosphate (5′-GMP) and 5′-inosine monophosphate (5′-IMP).

Production

The key factors in choosing the starting yeast for processing are the price, availability of the yeast itself, and the properties desired in the final product. Baker's, brewer's, and torula yeast strains serve as common substrates but in general, brewer's yeast requires a pretreatment prior to use. There are three manufacturing procedures for yeast extract production, namely, autolysis, plasmolysis, and hydrolysis (Hough and Maddox 1970).

Protein Hydrolysates

Protein hydrolysates have long been used for the nutritional management of people as well as for pets that cannot digest intact proteins. The most prevalent use of such hydrolysates is in the production of baby food formulas, particularly in foods for infants with hypersensitivity to food allergens. This is in part because allergic reactions to food occur most frequently with infants. Between 0.5% and 10% of normal full-term infants will develop some kind of food allergy, the most common being due to bovine milk because it is the most widely used source of protein for infants. (Kleinman et al. 1991) Since people and household pets exhibit similar food allergies, the pet food industry has adopted some of the same technologies as used in the food industry. Thus, as in case of many nutraceutical concepts which were proven in the food industry, the use of hydrolysates is also taking hold in pet food industry.

Food allergies are a serious health problem with symptoms ranging from mild skin rashes to life threatening conditions, such as anaphylactic shock. Many food allergies are transient in nature, and until recently, avoidance of the allergen (the allergy causing ingredient), and thus the entire allergenic food source, was the only available option to prevent serious allergic reactions. Researchers have found that the avoidance of a food allergen for a period of 1–2 years results in the loss of sensitivity to that allergen in approximately one-third of older children and adults allergic to foods (Sampson and Scanlon 1989; Pastorello et al. 1989). Additionally, other researchers have reported that as many as 85% of food-sensitive infants lose their sensitivity to those foods by 3 years of age. Unfortunately, this is not true of all allergens, as some allergenic foods have life-long allergic effects, including many that can be prolonged, permanent and life threatening (Cordle 1994).

As far as pets are concerned, however, no such data has been accumulated, or at least published, to the best of our knowledge. It is safe to assume though, that just as with people, if pets avoid food allergens for a long duration, they too will benefit from a loss of sensitivity to those allergenic food ingredients.

In the last several years, the need to control food allergies by methods other than avoidance has spurred the development of new technologies. One such development in the pet food industry has been the introduction of pet food formulas containing different sources of protein and the incorporation of hydrolysates of low molecular weight. Although immunoassay is considered the gold standard for evaluation of allergenicity, appropriate assays are not always available for specific hydrolysates. Fortunately, protein size can provide some indication of allergenic potential. As a rough rule of thumb, proteins smaller than 10,000 Da are weekly allergenic (immunogenic), and peptides less than 2,000 Da are generally not immunogenic without modification. Such hydrolysates from soy protein and casein are being utilized by the nutrition industry for geriatric and immuno-compromised populations.

Generally, protein hydrolysis can be achieved by either physical (heat and shear) or chemical (acid or enzymes) means. There are problems, however, with physical hydrolysis, as seen in the example of bovine milk. Milk is one of the most frequently implicated food groups in allergic reactions. It consists of a heterogeneous collection of 20–25 different proteins that significantly vary in their antigenicity. Additionally, milk from different animal sources may also vary in their intrinsic allergenic characteristics. For example, β-lactoglobulin (one of the two major whey proteins obtained from bovine milk) and casein (another major milk protein) are both allergenic. Casein, however, is unlike the majority of the milk proteins; it is extraordinarily stable and does not denature significantly at 130°C for 1 h. Most other soluble milk or whey proteins, such as β-lactoglobulin, denature at temperatures as low as 50–80°C. Thus, heat treating milk will certainly reduce the allergenicity to some of its proteins, but not all of them. Unfortunately, because the hydrolysis of all allergenic proteins is necessary to make milk tolerable to the entire allergenic population, physical means are often imperfect.

The other method of hydrolysis is chemical, which offers additional benefits that help combat some of the problems associated with physical hydrolysis. Chemical hydrolysis has often been used before. Acid hydrolysis of soy proteins, for example, is a well-established technology with its own merits. The use of enzymes however, is the preferred chemical method to hydrolyze proteins for food ingredients because enzymes also help in obtaining the desired lengths of peptides, an important factor in nutritional considerations.

Although all hydrolysis methods change the chemical, physical, biological, and immunological properties of the proteins (for example, enzyme hydrolysis changes solubility and emulsifying properties of proteins), the use of enzymes allows one to test a variety of available food grade options to correct for these changes to make them suitable for specific processing and nutritional needs (Mahmoud et al. 1992).

Researchers have demonstrated that low molecular weight peptides (di- or tri-peptides) with very low proportions of free amino acids are optimal for nutritional and therapeutic values. On the other hand, peptides with high molecular weights (containing 20 or so amino acids) improve the functionality and handling characteristics of the hydrolysates. Thus, the challenge for food technologists is to balance these competing concerns when selecting the molecular profile of any given hydrolysates.

The downside of using enzyme hydrolysis, however, is that proteolytic food grade enzymes not only alter the allergenicity and immunogenicity of proteins, but they also affect the taste of the food. Similarly, the initial choice of protein in the preparation of hydrolysates has a great effect on taste and mineral content, as well as process economics. In general, vegetable proteins are less expensive than animal proteins. Vegetable proteins, however, are not as good nutritionally as animal proteins, and tend to vary in amino acid content more than animal proteins, such as dairy proteins. Thus, in general, vegetable protein hydrolysates need fortification with amino acids and minerals.

Casein is one of the most desirable sources of protein in the preparation of hydrolysates because of its nutritional value, relative consistency, and economy. Unfortunately, it also poses a major palatability problem when hydrolyzed because it contains a higher than average proportion of hydrophobic amino acids. The bitter taste of these amino acids also causes the resulting hydrolysates to taste bitter. Additionally, casein also contains several highly acidic regions that contain phosphoserine sequences that are protease-resistant, and therefore reduce hydrolysate yields. Although de-bittering enzymes can be used to help alleviate this situation, such enzymes may reduce the nutritional value of the hydrolysate by generating less nutritionally desirable free amino acids from the smaller peptides. An alternate approach is to use a C-18 reverse phase resin to selectively remove hydrophobic peptides. Such peptides bind more firmly to the C-18 resin and are retained under conditions where more polar peptides do not interact. This approach was found to not only decrease the bitterness of the casein hydrolysates, but also to decrease antigenicity, as well. (Thomas et al. US Patent No. 5,837,312). It is because of these quality and economic concerns that casein hydrolysis has not been utilized in the pet food industry.

Enzymatic Hydrolysis of Proteins

As previously explained, enzyme hydrolysis offers several advantages, including: (1) controlled processing, (2) ease of obtaining a desired degree of hydrolysis, and (3) lack of necessity to deal with acid which would insert extra processing steps, such as conversion to a salt. The choice of a proteolytic enzyme to use in hydrolysis is crucial in determining the characteristics of the hydrolysates because it is the function of enzymes to break down large chains of proteins into small peptides and amino acids. Additionally, enzymes also constitute an important cost factor in determining the process economics. Depending on the desired degree of hydrolysis, today's food technologists/biochemists have a wide variety of food grade proteolytic enzymes to choose from within their economic constraints.

Proteases are often classified on the basis of the key amino acid involved in the active site mechanisms of the enzyme: serine; carboxyl; sulfhydryl; or metallo. The classification carries no implications regarding the specificity of a protease (e.g. trypsin, a serine protease, does not cleave proteins at serine residues), but a sound knowledge of classification can be useful in optimizing and troubleshooting the hydrolysis process. For instance, sulfhydryl proteases are generally inhibited by divalent metal ions and metallo proteases are inhibited by chelating agents such as EDTA and pyrophosphate.

Proteases are also sometimes classified by the relative pH range in which they are active (acid, neutral, or alkaline). Additionally, it is important to know whether an enzyme is an endoprotease (one that cuts between amino acids on the interior of the protein) or an exoprotease (which cleaves off amino acids at the ends of a protein or peptide). Exopeptidases can be either amino peptidases, which work at the amino end, or carboxypeptidases, which work at the carboxyl end.

Microbial rennets and rennin (chymosin) are examples of highly specific endoproteases. Pepsin has relatively broad specificity, but preferentially cleaves bonds involving Phe, Met, Leu, or Trp. On the other hand, fungal proteases are non-specific in nature and often exhibit both endo and exoprotease activities.

Alkaline proteases include subtilisin which is quite non-specific in nature. Neutral proteases are represented by trypsin which is a highly specific endoprotease that cleaves bonds between Arg or Lys and the amino acid that follows. Papain and bromelain are non-specific sulfhydryl endopeptidases. Several bacterial, neutral proteases are available and these, too, exhibit non-specific endoprotease properties.

Protein hydrolysis usually includes the use of endo as well as exoproteases to obtain optimal breakdown of the entire protein chain. The resulting hydrolysates contain free amino acids and small peptide chains. The use of different enzymes will change the peptide distribution and chemical properties of the resulting casein hydrolysates (Hudson 1995).

Enzymatic hydrolysis of proteins, however, does not always lead to a reduction in allergenicity. In fact, the initial breakdown of proteins exposes new antigenic epitopes, which actually results in increased antigenicity. As the hydrolysis progresses and the proteins are broken down to a greater degree, however, their antigenic properties lessen. Protein hydrolysates with an average molecular weight distribution lower than 800–1,500 Da are non-allergenic.

The breakdown of proteins and the resulting change in biological, physical, and chemical properties, also affect food processing, which may or may not be advantageous. In general, hydrolysis improves solubility as well as heat and pH resistance because of its low molecular weight distribution. On the other hand, it increases osmolarity and reduces chemical stability. Depending on the use of enzymes and the protein source, palatability also may be reduced due to the production of bitter peptides. Other effects of hydrolysis are hard to predict uniformly. For example, the absorption qualities and biological values of resulting hydrolysates will vary based on their respective protein sources. Similarly, as discussed before, depending on the degree of hydrolysis, allergenicity may be significantly changed or eliminated.

The processed food and flavor industries utilize highly hydrolyzed vegetable and animal proteins to create savory flavors. As previously discussed, the hydrolysis of the yeast protein may be accomplished by the manipulation of the endogenous yeast proteases or by supplementation with selected extraneous enzymes. Although these yeast peptide extracts have savory flavor characteristics, and are used without further processing, they can be made to undergo Maillard-type reactions to enhance/modify the savory notes.

An interesting parallel exists in the pet food industry, where a similar approach is taken in using protein hydrolysates to create palatability enhancers. These palatability enhancers are necessary because the typical pet food kibble is an extruded product that has only moderate inherent palatability. Many pets, and cats in particular, may even refuse to eat extruded pet foods that have not been treated with palatability enhancers. It is for this reason that such palatants are extremely important to pet food production.

Palatability enhancers come in both liquid and dry forms, are composed of a protein digest (usually meat), and are usually applied to the outer surface of the pet food, where it can have the greatest and most immediate impact on the pets' senses. The palatants are essentially a flavor concentrate, with a typical application rate of 1–3% for liquid versions and 1–2% for dry versions. These proportions can vary slightly because combinations of liquid and dry palatants may also be used.

This rapidly growing industry, however, has an even greater series of economic constraints than the human food production industry. These economic constraints and the fact that visceral by-products are desired by pets, means that liver or viscera of poultry and animal by-products are often the protein of choice in pet food production. One advantage with viscera is that it has an abundance of proteolytic enzymes. Thus, it requires very little, if any, additional protease in order to hydrolyze proteins. The hydrolysates employed by general pet food products are of a lower degree of hydrolysis. Although this offers many processing advantages, the molecular weight distribution of these hydrolysates is not in the range of rendering them non-allergenic.

When using certain animal organs as protein sources, it is possible to take advantage of endogenous proteolytic enzymes, although often these enzymes have too high a representation of exoproteases to produce a good digest. If necessary, the meat source can be ground and treated with enzymes selected to provide a high yield of peptides with savory characteristics. For obvious reasons, a broad-specificity endoprotease is usually chosen to do this. Examples of commercially available enzymes include the sulfhydryl protease, papain, and a serine protease of bacterial origin. Enzymes with more restricted specificity may also be used to produce digests with flavor characteristics that appeal to a targeted pet population. Because the flavor preferences of cats and dogs do not tend to coincide, pet food manufacturers try to select enzymes that create digests with high yields and appropriate flavor characteristics.

As in the human food industry, in addition to maximizing the palatability of the resulting peptides, it is also desirable that the chosen proteases lend themselves to 'friendly' processing conditions, e.g. pH and temperature ranges that help exclude bacterial growth.

As in the case of yeast extracts mentioned above, sugars and other ingredients can be added to the digest to allow it to undergo a Maillard-type reaction. The resulting reaction product could be combined with other palatability-enhancing ingredients and be used directly in the pet food as either a liquid or spray-dried flavoring, or it may be subsequently blended with other ingredients to make a dry palatant.

The technical challenge for food technologists and biochemists is to use an appropriate combination of enzymes within the existing economic constraints and other physical factors/limitations, such as heat, pH, and time, to create highly palatable, yet still nutritious and hypoallergenic food formulations. Developments in enzyme engineering in the years to come should help food technologists and biochemists to design enzymes that cleave resistant amino acid sequences to yield protein hydrolysates that are increasingly hypoallergenic to the point of virtually eliminating the risk of food allergies in specialized formulations, while still maintaining process economics.

Composition of Protein Hydrolysates

It is always difficult to know the composition of protein hydrolysates because of the multi-degree polymerization of peptides, i.e. peptides that are characterized by 20 natural amino acids. In the human and pet food industries this is over-come to some degree by employing multiple analyses for quality assurance and quality control purposes. Mainly, these analyses measure the degree of hydrolysis represented by the: (1) ratio of amino nitrogen over total nitrogen (AN/TN Ratio) in the resulting hydrolysate; (2) presence of amines in the hydrolysate; and (3) osmolarity of the hydrolysate.

Degree of Hydrolysis

There are many different methods used in the food and pet food industry to quickly determine the degree of hydrolysis (DH) of the peptide bonds of food proteins. One would assume that at 25% DH, every fourth peptide linkage would be broken down, creating mostly tetrapeptides, or at least hydrolysates that only have very small molecular weights. This simple assumption, however, is wrong because the breakage randomly results in rather uneven lengths of peptides with an average length of four amino acids.

Although the parent proteins themselves have a defined amino acid sequence, the non-specificity of the endoproteases can lead to heterogeneity in the peptide released. For instance, if two potential cleavage sites in a protein exist in close proximity to each other, then initial cleavage at one of the sites may preclude cleavage at the other site because the activity of endoproteases, in general, is influenced by the presence (or absence), as well as the identity, of the amino acids around the cleavage sites.

This is compounded by a degree of randomness in the denaturation pattern of proteins as they undergo hydrolysis, which exists even under controlled process conditions. This randomness influences the selection of cleavage sites, as described above. Thus, even under identical conditions, one does not obtain exactly the same peptide distribution in a hydrolysate from one time to the next. In practice, these differences are normally of slight importance in the properties of the resulting hydrolysates, although it could conceivably contribute to batch-to-batch differences in bitterness and/or allergenicity.

The methods used in the quantitative determination of the degree of hydrolysates employ one of the following principles: (1) determination of soluble nitrogen in the presence of a precipitating agent such as trichloro acetic acid (TCA); or (2) determination of free alpha amino groups by colorometric methods (e.g. titration with trinitrobenzenesulfonic acid, TNBS), or pH titration of the released protons. TCA-soluble nitrogen may be determined by the Kjeldhal assay (A.O.A.C 1995) or the Biuret reaction (Hung et al. 1984). Silvestre (1997) has described these methods in detail.

Qualitative and Quantitative Analysis

In order to better understand the nutritional value and the characteristics of the hydrolysates, it is important to know their chemical compositions. Reverse Phase Mode High Performance Liquid Chromatography (RP-HPLC) efficiently separates peptides from protein hydrolysates, and also provides some indication as to their hydrophobicity and hydrophilicity. Several researchers have employed this technique in studying tryptic hydrolysate of casein from β-lactoglobulin A (Kalghatgi and Horvath 1987; Maa and Horvath 1988). Isolation of bitter peptides from the enzymatic hydrolysates can also be carried out by RP-HPLC. In 1989, Minagawa and his colleagues used this technique for isolating the bitter peptide fraction from a tryptic casein hydrolysate (Minagawa, et al. 1989). Based on chromatographic and electrophoretic behavior, several other techniques have been developed and used to separate and identify peptides and amino acids in protein hydrolysates. However, the problems related to the interactions between the solutes and the matrix has not yet been completely resolved.

A common method for estimation of the molecular weight distribution of hydrolysates is size exclusion high pressure liquid chromatography (SEC-HPLC). Such chromatography media separates on the basis of size. By calibrating the column with peptides/proteins of known molecular weight, the molecular weight distribution of the peptides in a hydrolysate can be estimated.

Protein hydrolysates are very complex, usually containing mixtures of many peptides of similar size as well as amino acids; this complicates quantitation of the protein content of fractions from chromatography columns. The challenges have been reviewed (Silvestre et al. 1994) and a method based on UV absorbance measurement at 230 nm with correction for the aromatic amino acids absorbance using multi-detections at 230, 280 and 300 nm wavelengths was suggested.

Polyacrylamide gel electrophoresis in the presence of sodium dodecyl sulfate (SDS-PAGE), coupled with silver staining is also a useful indicator of molecular weight. Mass spectroscopy, used alone and in-line with HPLC offers an alternate method for characterizing hydrolysate molecular weight distributions.

In summary, the characterization of protein hydrolysates may be done by the evaluation of the degree of hydrolysis of the peptide bonds as described previously. Even though the methodology involved seems quite simple, there are several interfering factors that make data interpretation difficult and problematic. Different techniques based on chromatographic or electrophoretic behavior have been used to separate and identify amino acids and peptides in the hydrolysates of animal and vegetable proteins for quantitative analysis. The estimation of the molecular weight distribution of the resulting hydrolysis is a very important analysis. This is also determined by chromatographic or electrophoretic techniques. UV spectrophotometry has also been used to analyze hydrolysates. It can give a rapid estimation of the degree of hydrolysis and homogeneity of the hydrolysates.

Future Trends

The Western World, which now produces and consumes vast quantities of animal meat may, in the future experience shortages that could result in dramatic changes in food habits. Additionally, consumers have now become exceedingly conscious of their health and this too has caused a dramatic reduction in the inclusion of meat in their daily diet. In response to these pressures and consumer demands, meat analogs are becoming increasingly popular, and as this occurs, production of meat-like savory flavors will be of ever increasing importance throughout the world.

It is the responsibility of the food scientists, flavorists and product designers to develop savory richness in diets low in meat or in those that are rich in vegetable proteins. Currently, there is considerable interest to produce very high 5-nucleotide-rich yeast extracts for use in flavor enhancement.

All the mechanisms associated with the Maillard reaction as regards to flavor generation are obviously very complex. Many pathways have been proposed but it is still necessary to investigate further to thoroughly understand it. Current research is devoted more to the identification of compounds contributing to savory flavor and to maximize the production of these flavor chemicals. Understanding the kinetics of these reactions would be helpful to activate only the relevant pathways that would generate the desired flavors without the release of undesirable flavors. In the future, we can expect to see further interest in model systems for both knowledge building as well as for practical use.

The quest for controlled flavor release will continue to remain a high priority to those who are in the business of formulating prepared foods. The basic intention is to provide the fresh flavor notes that are reminiscent of fresh homemade foods. One approach that has not yet reached perfection is to add flavor precursors, which may include specially designed yeast extracts and protein hydrolysates, in addition to flavorings. The product designers are aggressively making headway in taking advantage of this approach. The hope is that such well-defined flavor precursors introduced into the formulation would generate the characteristic taste and aroma during the final heat treatment, just prior to consumption to make the food resemble freshly prepared foods.

As described earlier, the antigenicity of dietary proteins can be minimized or perhaps eliminated by enzymatic hydrolysis to produce low molecular weight peptides that would make the product hypoallergenic. The future in this area would thus be to develop such hydrolysates to have, in addition to hypoallergenicity, excellent solubility and high heat stability to make them suitable for use in medical foods and sports nutrition where such products are easier to consume and absorb. In such beverage-type applications, the hydrolysates must secure the right taste and functionality for easy application.

A few well-known food and pet food research laboratories have come closer than ever before in acquiring control over the process of achieving flavor development through the application of model systems. These developments have set the stage for flavor chemists to utilize the power of computers to assist in steering reactions to generate definite flavors, important for new product development. This will allow food and pet food research to gradually move away from the "hit-and-miss" type approach to a more defined and predictable flavor design and flavor engineering approach in the future.

References

A.O.A.C. (1995) Official methods of analysis, 16th edn, Ch. 12. Horowitz, Washington, DC, pp 7–9

Cordle CT (Oct 1994) Control of food allergies using protein Hydrolysates. Food Technol 10:72–76

Heer J (1991) Nestle 125 years, 1866-1991. Nestle SA Veve, Switzerland

Hesseltine CW (1983) Microbiology of oriental fermented foods. In: Ornston LN (ed) Annual reviews of microbiology, vol 37. Annual Reviews Inc., Palo Alto, CA, pp 57–64

Hough JS, Maddox IS (1970) Yeast autolysis. Proc Biochem 5:50–57

Hudson MJ (1995) Product development horizons- a review from industry. Eur J Clin Nutr 49(Suppl 1):S64–S70

Hung ND, Vas M, Cheke E, Bolesi SZA (1984) Relative tryptic digestion rates of food proteins. J Food Sci 49:1535–1542

Kalghatgi K, Horvath C (1987) Rapid analysis of proteins and peptides by reverse phase chromatography. J Chromatogr 398:335–339

Kleinman RE, Bahna S, Powell GF, Sampson HA (1991) Use of infant formuals in infants with cow milk allergy. A review and recommendations. Pediatr Allergy Immunol 4:146–155

Maa YF, Horvath C (1988) Rapid analysis of proteins and peptides by reverse phase chromatography. J Chromatogr 398:335–339

Mahmoud MI, Malone WT, Cordle CT (1992) Enzymic hydrolysis of casein: effect of degree of hydrolysis on antigenicity and physical properties. J Food Sci 57:1223–1229

May CG (1991) Process flavorings. In: Ashurst PR (ed) Food flavorings. Van Nostrand Reinhold, New York, pp 257–286

Minagawa E, Kaminogawa S, Tsukasaki F, Uamauchi K (1989) Debittering mechanisms in bitter peptides of enzymic hydrolysates from milk casein by aminopeptidase T. J Food Sci 54:1225–1229

Nagodawithana TW (1994) Savory flavors. In: Gabelman A (ed) Bioprocess, production of flavors, fragrances and color development. Wiley, New York, pp 135–168

Nagodawithana TW (1995) Maillard and other flavor-producing reactions. In: Nagodawithana TW (ed) Savory flavors. Esteekay, Milwaukee, WI, pp 103–163

Pastorello EA, Stocchi I, Pravetonni V, Bigi A, Schilke MI, In Cuvaria C, Zanussi C (1989) Role of the elimination diet in adults with food allergy. J Allergy Clin Immunol 84:475–483

Prinsen-Geerligs HC (1896) Einige Chinesische Sojabohnenprraparate Chem. Zeit 20:67–73

Reed G, Nagodawithana TW (1991) Yeast derived products. In: Nagodawithana TW (ed) Yeast technology. Van Nostrand Reinhold, New York, pp 369–412

Sampson HA, Scanlon SM (1989) Natural history of food hypersensitivity in children with atopic dermatitis. J Pediatr 11:23–27

Silvestre MPC (1997) Review of methods for the analysis of protein hydrolysates. Food Chem 60:263–273

Silvestre MPC, Hamon M, Yuon M (1994) Analysis of protein hydrolysates. 2. Characterization of casein hydrolysates by a rapid peptide quantification method. J Agric Food Chem 42:2783–2789

Steinkraus KH (1983) Indigenous fermented amino acid/peptide sauces and pastes with meat-like flavor. In: Steinkraus KH (ed) Handbook of indigenous foods, vol 9, Microbiology series. Marcel Dekker, New York, pp 433–468

Thomas CC, Lin S, Nelles N Debittered protein product having improved antigenicity and method for manufacture. U.S. Patent No. 5,837, 312

Yokotsuka T (1972) Some recent technological problems related to the quality of Japanese shoyu. In: Terui G (ed) Fermentation technology today. Proceedings of fourth international fermentation symposium, Kyoto, Japan, March 19

Yokotsuko T (1986) Soy sauce biochemistry. In: Chichester CO, Mrak EM, Schweigert BC (eds) Advances in food research, vol 30. Academic, New York, p 195

Chapter 12
The Development of Novel Recombinant Human Gelatins as Replacements for Animal-Derived Gelatin in Pharmaceutical Applications

David Olsen, Robert Chang, Kim E. Williams, and James W. Polarek

Abstract We have developed a recombinant expression system to produce a series of novel recombinant human gelatins that can substitute for animal sourced gelatin preparations currently used in pharmaceutical and nutraceutical applications. This system allows the production of human sequence gelatins, or, if desired, gelatins from any other species depending on the availability of the cloned gene. The gelatins produced with this recombinant system are of defined molecular weight, unlike the animal-sourced gelatins, which consist of numerous polypeptides of varying size. The fermentation and purification process used to prepare these recombinant gelatins does not use any human- or animal-derived components and thus this recombinant material should be free from viruses and agents that cause transmissible spongiform encephalopathies. The recombinant gelatins exhibit lot-to-lot reproducibility and we have performed extensive analytical testing on them. We have demonstrated the utility of these novel gelatins as biological stabilizers and plasma expanders, and we have shown they possess qualities that are important in applications where gel formation is critical. Finally, we provide examples of how our system allows the engineering of these recombinant gelatins to optimize the production process.

Keywords Gelatin • *Pichia pastoris* • Spongiform encephalopathies • Recombinant and molecular weights

Introduction

Gelatin is prepared by a multi-step process using hides or bones usually from cows or pigs. More recently gelatin preparations from fish and poultry have been introduced. Tissues from numerous animals are pooled and the fat and mineral components are removed. The material is put through either an acid (type A gelatin) or alkaline (type B gelatin) process, depending on the type of gelatin to be produced.

D. Olsen (✉), R. Chang, K.E. Williams, and J.W. Polarek
FibroGen, Inc., 409 Illinois Street, San Francisco, CA 94158, USA
e-mail: dolsen@fibrogen.com

V.K. Pasupuleti and A.L. Demain (eds.), *Protein Hydrolysates in Biotechnology*,
DOI 10.1007/978-1-4020-6674-0_12,

In the alkaline process the material is treated with lime for several months, rinsed extensively and the gelatin is extracted with warm water and separated from the other components. The acid process is more rapid in that the raw material is acid treated for only 1 day, neutralized, rinsed extensively and then extracted with warm water. Since these gelatin preparations are made from pooled tissues it is difficult to trace the exact source of each preparation. Also these gelatins can exhibit batch-to-batch heterogeneity. Only minimal analytical characterization can be performed on animal derived gelatin due to its heterogeneous nature.

Gelatin is used in numerous pharmaceutical applications including excipients used to stabilize vaccines and biologics, plasma volume expanders to treat hypovolemic shock, coatings for medical devices and microcarrier beads for cell culture, and it is the main component of hard and soft gel capsules. The type of gelatin used for these different pharmaceutical applications requires very different physical properties and thus additional specialized processing, beyond the type A or B process, may be required for gelatin in certain products. These steps include hydrolysis, chemical modification of amino acids, and cross-linking. The recombinant system we have developed to produce gelatin replacements allows us to manufacture recombinant human gelatins that can be tailored and optimized for each of the varied indications described above and will not require further specialized process steps. The process can be run under GMP conditions and yields highly purified, consistent lots of material. The majority of our efforts have focused on expression of fragments of the human $\alpha 1$(I) chain of type I collagen. We have also expressed gelatins encoding part or all of the sequences from human $\alpha 2$ (I), $\alpha 1$ (II) and $\alpha 1$ (III) chains but this discussion will focus on expression of gelatins encoded by the human $\alpha 1$(I) chain.

We have selected *Pichia pastoris* as the host organism to express these recombinant human gelatins. *Pichia* has been shown to be a very suitable strain for recombinant gelatin production (Olsen et al. 2005, 2003; Werten et al. 1999; Werten et al. 2001). This organism can be fermented to very high cell density in completely defined media, allowing maximal product accumulation in the absence of any added animal protein sources. We have evaluated both inducible and constitutive promoters and the generation of multi-copy strains to optimize production. Expression and secretion of a number of recombinant human gelatins, ranging in size from 56 to 1,014 amino acids has been demonstrated (Olsen et al. 2003). The properties and utility of a subset of these gelatins will be discussed.

Results and Discussion

Expression Optimization in Pichia pastoris

Pichia has been shown to be a highly productive system for expression of numerous recombinant proteins (Cregg et al. 1993). As is the case with other expression systems, some proteins undergo proteolysis during expression in

Pichia (Boehm et al. 1999; Brierley 1998; Clare et al. 1991; Werten et al. 1999; White et al. 1995). Several strategies have been tested to control proteolysis including optimization of fermentation parameters (temperature, pH), addition of supplements such as casamino acids to the fermentation broth, use of protease deficient host strains, and use of different promoters to drive expression (Goodrick et al. 2001; Werten et al. 1999). Our goal was to develop a highly productive, cost efficient process, devoid of any animal-derived components, that yields consistent material. A description of the recombinant human gelatins to be discussed can be found in Table 1.

Gelatin is distinct from collagen, the protein from which it is derived, in that it does not have a triple helical structure. The triple helix of collagen makes the protein resistant to proteolytic attack (Bruckner and Prockop 1981). Gelatins, by nature of their denatured form, are very sensitive to protease attack. Consequently, the documentation of proteolysis of animal sequence gelatins expressed in *Pichia* is not surprising (Werten et al. 1999). Proteolysis of mouse and rat gelatins was reduced to some extent by optimizing fermentation conditions but was not completely eliminated.

We have observed similar proteolytic processing of human gelatins expressed in *Pichia.* We were able to reduce proteolysis to some extent by utilizing protease deficient host strains and running the fermentation process at low pH to minimize the activity of yeast proteases. However, as was the case with animal sequence gelatins, proteolysis could not be completely eliminated. We have not considered the use of casamino acid supplementation to reduce degradation, as has been reported (Clare et al. 1991), since this supplement is derived from an animal protein (casein) and is thus undesirable.

To effectively control proteolysis we have used protein engineering to eliminate protease cleavage sites that were identified in recombinant gelatins following expression and purification. An example of the use of protein engineering to eliminate proteolytic processing of a 100 kDa human gelatin (FG-5012) is shown in Fig. 1. Following expression of the wild-type 100 kDa gelatin in *Pichia* we identified two major protease cleavage sites. Both cleavage sites were on the C-terminal side of arginine residues. The amino acid proline was substituted for the arginines at both of the identified cleavage sites. Expression of this engineered molecule, FG-5020, resulted in a significant reduction in product proteolysis during expression.

Table 1 Description of recombinant human gelatins

Designation	Prepro α1(I) amino acid #	Molecular weight (kDa)
FG-5009	531–630[a]	8.5
FG-5010	531–781	25
FG-5011	531–1030	50
FG-5019	531–1192	65
FG-5012	178–1192	100
FG-5020	178–1192[b]	100

[a]Six amino acid substitutions were introduced at positions 541, 543, 546, 553, 580, and 581 (Olsen et al. 2005).

[b]Proline residues were substituted for arginine residues at positions 220 and 304.

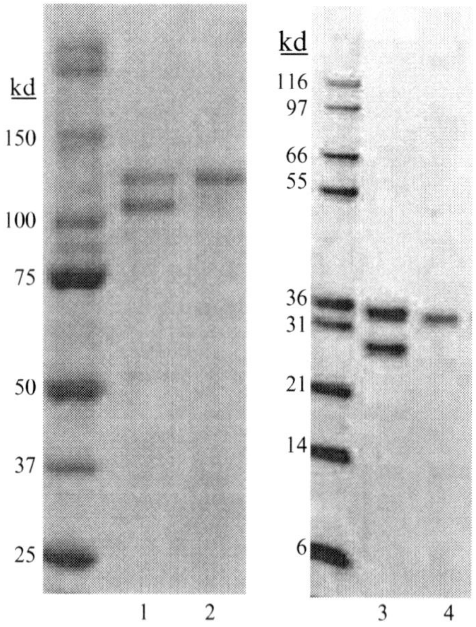

Fig. 1 Optimization of recombinant human gelatin expression. The wild-type (FG-5012; lane 1) and engineered (FG-5020; lane 2) 100 kDa human gelatins were expressed in *Pichia pastoris* and the cell-free conditioned media was analyzed by SDS-PAGE. A 25 kDa gelatin (FG-5010) was expressed in *Pichia pastoris* under the control of the AOX1 inducible (lane 3) or GAP constitutive (lane 4) promoter and the cell-free broth was analysed by SDS-PAGE

This result is similar to our previous findings where we introduced proline residues at a proteolytic cleavage site in an 8.5 kDa human gelatin and reduced proteolysis considerably (Olsen et al. 2005).

The use of a constitutive promoter to derive heterologous protein expression has been shown to be an alternative means of controlling proteolysis in *Pichia pastoris*. Expression of human chitinase was evaluated using a protease deficient *Pichia pastoris* strain (SMD1168; *pep4*) and a methanol fed-batch fermentation strategy. Degradation of the product was observed after 4 days of fermentation, even in the protease deficient background. A strain that expressed the same protein using the constitutive glyceraldehyde phosphate dehydrogenase (GAP) promoter yielded intact enzyme and the fermentation process could be run for 30 days (Goodrick et al. 2001). This same strategy was tested with a 25 kDa recombinant gelatin (FG-5010), derived from the human $\alpha 1(I)$ cDNA by PCR and cloned into expression vectors using both the methanol inducible AOX1 and constitutive GAP promoters. The constructs were electroporated into wild-type *Pichia* strain X-33 and strains expressing and secreting this gelatin were identified. A significant improvement in the quality of FG-5010 was noted when it was expressed using the GAP promoter (Fig. 1). Similar improvements in product quality were observed with other human gelatins (data not shown).

In addition to the improved quality of proteins produced in *Pichia* using the constitutive promoter there are other advantages to this approach. Several investigators have demonstrated yield improvements using a continuous, constitutive fermentation process to express hepatitis B surface antigen, human angiostatin, and chitinase (Goodrick et al. 2001; Vassileva et al. 2001; Zhang et al. 2004). There are also safety and cost advantages to using constitutive strains and glucose as a carbon source to express recombinant proteins at large scale. First, these strains/processes avoid the use of methanol, a more costly carbon source than glucose. Secondly, this eliminates a flammable solvent from the process and the need for an explosion-proof production plant.

We have exemplified several options for improving the quality of human gelatins expressed in *Pichia pastoris*. Several different molecular weight recombinant human gelatins were expressed, purified, and tested in various biological systems and models to demonstrate their utility. Three different molecular weight gelatins will be discussed in detail.

Low Molecular Weight Non-gelling Human Gelatins for Stabilization of Pharmaceuticals

Several live-attenuated viral vaccine formulations contain gelatin as an essential stabilizing component of the final formulation (Burke et al. 1999; deRizzo et al. 1989). The gelatin acts to protect the virus during the freezing and lyophilization process and during extended storage of liquid formulations. The gelatin used for this application must remain a liquid at cold temperatures and not form a solid gel. This non-gelling phenotype is achieved by hydrolyzing a high molecular weight gelatin preparation to one that has an average molecular weight in the range of 1–30 kDa. Alternatively, non-gelling gelatin preparations have been made by enzymatic treatment (Sakai et al. 1998). We have produced a non-gelling human gelatin, FG-5009, that remains a liquid in formulations at 2–8°C. FG-5009 has a molecular weight of 8.5 kDa and an isoelectric point (pI) of 8.4. FG-5009, unlike the animal-derived gelatin hydrolysates, does not contain a wide distribution of different molecular weight species, but rather consists of a single molecular weight polypeptide. This is shown by SDS-PAGE (Fig. 2) and also by size exclusion chromatography analysis (Fig. 3).

We have shown that a low molecular weight human gelatin of defined molecular weight stabilizes a live attenuated viral vaccines as well as animal gelatin hydrolysate (Olsen et al. 2005). FG-5009 can be made in a reproducible manner using a controlled fermentation process and it has been characterized extensively using a battery of analytical tests. In addition to its defined molecular weight and pI, FG-5009 also contains very low endotoxin and bioburden levels (Table 2). These purity features, not obtainable with animal-sourced gelatin, allow the use of greater quantities of material in formulations without concerns about unacceptable levels of these types of contaminants in the final product.

An additional concern associated with animal sequence gelatins used in vaccine formulations is allergic reactions. The types of allergic reactions documented range from non-immediate to immediate type reactions, including anaphylaxis (Sakaguchi and Inouye 2000). The sera from children with immediate type reactions to gelatin

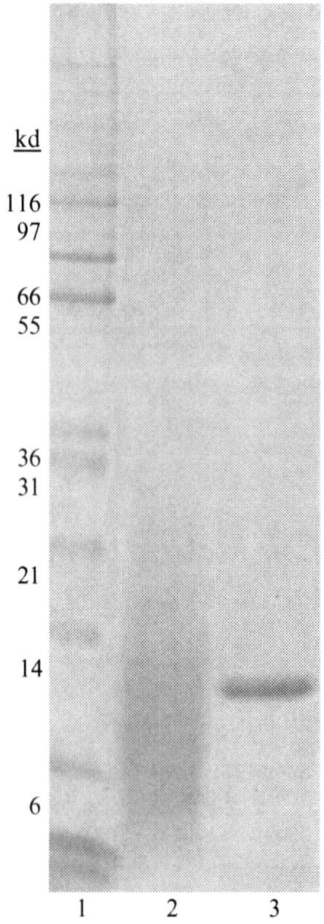

Fig. 2 Comparison of gelatin molecular weight distribution by SDS-PAGE. The 8.5 kDa recombinant human gelatin FG-5009 (lane 3) expressed in *Pichia pastoris* was analyzed on a 10–20% Tricine gel and stained with Gelcode Blue. A porcine gelatin hydrolysate (K&K) was run in lane 2 for comparison and molecular weight markers are shown in lane 1

contained anti-gelatin IgE antibodies (Sakaguchi et al. 1995). The epitope recognized by these IgE antibodies was shown to reside on the α2 chain of type I collagen (Sakaguchi et al. 1999). These antigenic epitopes are not present in FG-5009 since this protein is encoded by sequences from the human α1(I) cDNA. We have also directly demonstrated, using a binding assay, that anti-gelatin IgE antibodies from the sera of these sensitized children do not recognize FG-5009 (Olsen et al. 2005). Thus, FG-5009 has a superior immunologic safety profile compared to animal-sequence derived gelatin stabilizers.

The ability to perform extensive analytical testing on pharmaceutical formulations is advantageous during preformulation studies, formulation design and formulation process development. The inclusion of a protein excipient, such as serum albumin, can interfere with certain spectroscopic measurements used during stability testing and formulation development. Recombinant human gelatins, specifically FG-5009, have a distinct advantage since their primary amino acid sequence lacks tryptophan and tyrosine residues, amino acids that absorb in the 260–280 nm UV range. Stability indicating assays or release assays that measure absorbance in the 260–280 nm range can be performed on formulations containing FG-5009 as a stabilizer without any interference due its unique amino acid composition. This is demonstrated in Fig. 4, which shows the results of a cation exchange chromatographic analysis of FG-5009 monitored at 215 and 280 nm.

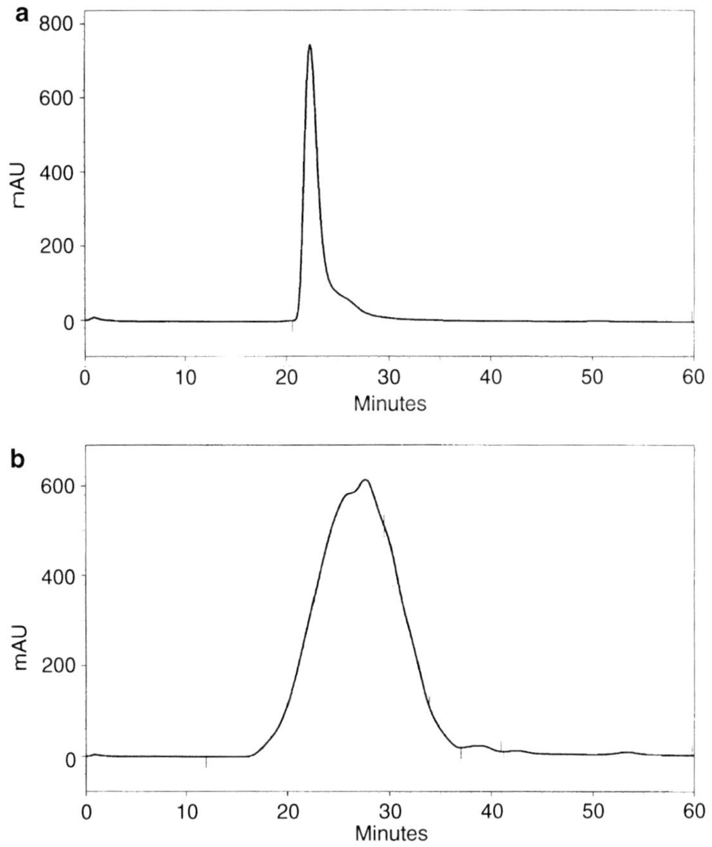

Fig. 3 Analysis of gelatin molecular weight distribution by size exclusion chromatography. Approximately 50 μg of FG-5009 (Panel A) or K&K porcine gelatin hydrolysate (Panel B) were analyzed by size exclusion chromatography on a Superdex 75 column in 100 mM sodium phosphate buffer pH 6.0. Absorbance was monitored 215 nm

Table 2 List of specifications for FG-5009

Analytical assay	Specification
Appearance	White solid, free from visible contamination
pH	6.0–9.5
SDS-PAGE	≥95% total gelatin, no single impurity ≥0.5%
Protein content	80.0–105%
Host cell Protein	≤0.05 ppm (L.O.D.)
Host cell DNA	≤5 pg/mg gelatin
Residual carbohydrate	≤2 μg/mg gelatin
Bioburden	≤20 cfu
Endotoxin	≤0.1 EU/mg gelatin
N-terminal sequence	Matches reference standard
Heavy metals	≤50 ppm

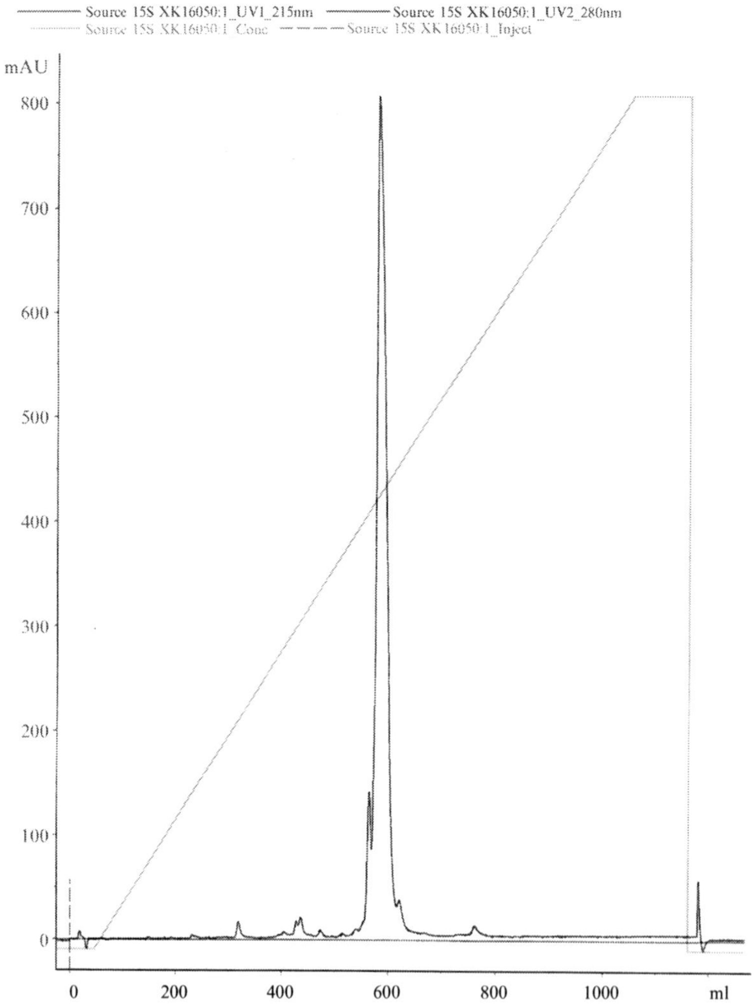

Fig. 4 Analysis of FG-5009 by cation exchange chromatography. Five milligrams of FG-5009 were fractionated on a Source 15S column and monitored at 215 nm (*blue trace*) or 280 nm (*red trace*). FG-5009 does not shown any significant absorbance at 280 nm due to the lack of aromatic amino acids

Recombinant Human Gelatins for Use as a Plasma Expander

Animal sourced gelatin formulations for treatment of hypovolemic shock have been available in Europe for many years. The gelatins used for this indication have average molecular weights of about 25 kDa and are either chemical cross-linked or succinylated prior to use to enhance plasma half-life (Saddler and Horsey 1987). The animal source and heterogeneity of the gelatin used to formulate these products are limitations. These preparations contain a distribution of different size molecules and a portion of the lower molecular weight species are rapidly cleared by the kidneys, limiting therapeutic effectiveness. Also, anaphylactic reactions have been associated with their use (Vervloet et al. 1983).

We have expressed 50 and 65 kDa recombinant human gelatins (FG-5011 and FG-5019, respectively) for use as plasma expanders. These molecular weight species were chosen since they should not be rapidly cleared by glomerular filtration in the kidneys. These materials were compared to Gelafundin, a 4% solution of succinylated bovine gelatin marketed in Europe as a plasma expander. The molecular weight distribution of the three preparations was compared by SDS-PAGE analysis (Fig. 5). The Gelafundin is heterogeneous, consisting of many different sized polypeptides while the recombinant gelatins consist of a single molecular weight species. FG-5011 and FG-5019 were also succinylated in the same manner as Gelafundin prior to testing (Tourtellotte and Williams 1958).

A canine model of hypovolemic shock was established in which 40% of the blood volume was removed from 10 to 12 kg beagle dogs, previously implanted with radiotelemetry devises to monitor blood pressure. Following exfusion, the mean arterial blood pressure (MAP) dropped significantly to ~30 mmHg. The volume of blood exfused was replaced by intravenous infusion of FG-5011 or

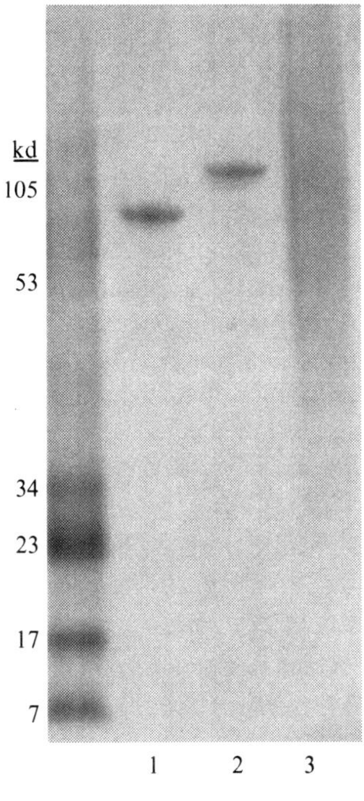

Fig. 5 Comparison of molecular weight distribution of plasma expander formulations. The recombinant gelatins and Gelafundin tested as plasma expanders were analyzed on 10–20% Tricine gels and proteins were stained with Gelcode Blue. Lane 1, 5 μg 50 kDa recombinant human gelatin FG-5011; Lane 2, 5 μg 65 kDa recombinant human gelatin FG-5019; Lane 3, 30 μg Gelafundin

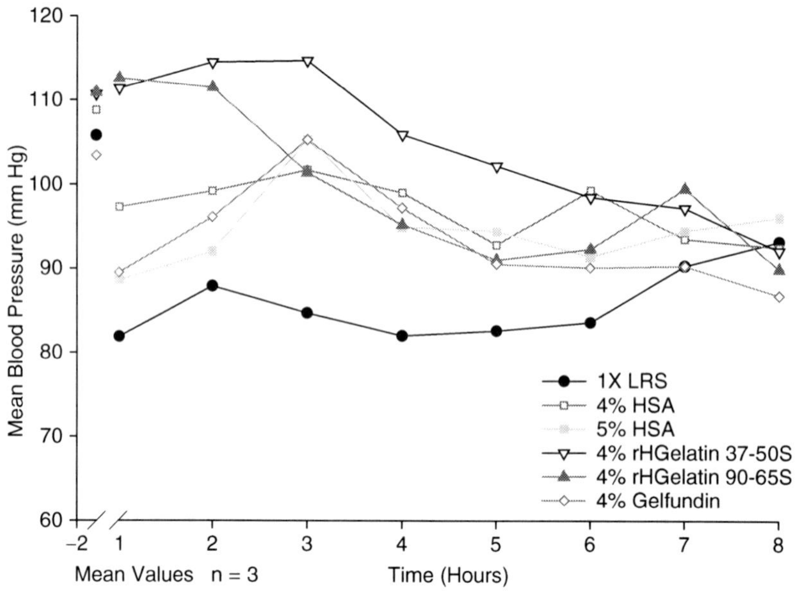

Fig. 6 Comparison of plasma expander activity of various formulations in a canine model of hypovolemic shock

FG-5019 formulated in Lactated Ringers solution (LRS). In addition to these two test articles, LRS, human serum albumin (HSA) or Gelafundin (both formulated in LRS) were tested in parallel. The recombinant gelatins, especially FG-5011, induced a rapid and prolonged restoration of MAP. The kinetics and duration of the effect with the 50 kDa gelatin was better than that seen with HSA, Gelafundin, or LRS (Fig. 6). These data demonstrate that FG-5011 is a very effective plasma expander, inducing a rapid and sustained recovery of blood pressure in this model of hypovolemic shock. Additionally, the recombinant gelatins contain no animal-derived components and are homogeneous preparations (Fig. 5).

FG-5011 has also been used to coat microcarriers to promote the attachment of several mammalian cells, including Vero cells (Olsen et al. 2003). Vero cells are grown on microcarriers in large bioreactors to produce viral stocks for vaccine production (Montagnon et al. 1983). Our attachment studies were performed in serum-free media and demonstrate that our recombinant human gelatin can substitute for the animal gelatin that is currently used to coat microcarriers.

High Molecular Weight Gelatins with Gel Forming Properties

Gelatin has been used in the food and pharmaceutical industry for many applications for almost a century. The gelatin used for theses applications has the ability to reversibly form a solid gel upon heating and cooling. The production of hard and soft shell gelatin capsules for drug delivery relies on this property. In some applications the gelatin adds texture, viscosity, and the ability to create foam in certain products. Additionally, gelatin used in confectionery and dairy products melts slowly in the mouth creating a smooth taste sensation, gently releasing flavor. The gelatin preparations used for these applications have physical properties that are very different from the gelatins used as excipients to stabilize pharmaceuticals and plasma expanders. In order to form a solid gel of appropriate strength they must have a certain Bloom strength, viscosity, set rate, film forming property, and tensile strength.

Both type A and type B gelatins or blends of these two types are used in applications where gelling properties are important. These preparations typically contain a mixture of full-length collagen alpha chains, cross-linked species called β or γ components and varying amounts of degraded material. The ratio of these different components affects Bloom strength and viscosity and thus gel formation. The source of the tissue used for gelatin isolation (bone or hide), the type of extraction process, and the age of the tissue used all influence the molecular weight distribution of the gelatin obtained from the process. All of these variables make the reproducible manufacture of an ideal gelatin a challenge.

In addition to these parameters, other challenges are presented in specific applications, for example, the gelatin used to manufacture capsules. One of the limitations of current gelatin capsules is a reduction in the rate of dissolution, thought to be the result of cross-linking reactions occurring in the outer layer of capsules (Carstensen and Rhodes 1993; Mhatre et al. 1997; Ofner et al. 2001). These cross-links involve lysine and/or arginine residues in gelatin (Gold et al. 1996; Ofner and Bubnis 1996). An additional concern is the interaction of the active pharmaceutical ingredient with the capsule affecting bioavailability (Gholap and Singh 2004).

In order to produce a replacement material for gelatins with gelling properties we initially expressed a recombinant human gelatin similar in molecular weight to the full-length α chain component found in animal derived gelatin. Due to the nature of our recombinant system, this 100 kDa gelatin (FG-5012) does not contain β or γ components and yields a homogeneous, more reproducible material. During the early development of this recombinant gelatin, proteolysis led to fragmentation of the material during expression in *Pichia* (Fig. 1). We have since developed an engineered version of this gelatin, FG-5020, which is protease resistant and results in the production of higher quality material. The homogeneity of this recombinant gelatin was compared to several tissue-derived gelatin preparations that form solid gels upon cooling. The results of these comparisons are shown in Figs. 7 and 8. The results clearly demonstrate the homogeneous nature of FG-5020, as compared to the heterogeneous tissue derived gelatins.

We have studied some of the physical properties of FG-5020 including Bloom strength, viscosity, thermoviscosity breakdown and structural conformation. The results of the Bloom and viscosity analyses are summarized in Table 3. For comparison, these same properties were measured with an animal-derived type B gelatin. FG-5020 has a Bloom value of 225 g and a viscosity of 3.69 cP at 60°C, similar to the bovine type B gelatin control. These Bloom and viscosity values are in the

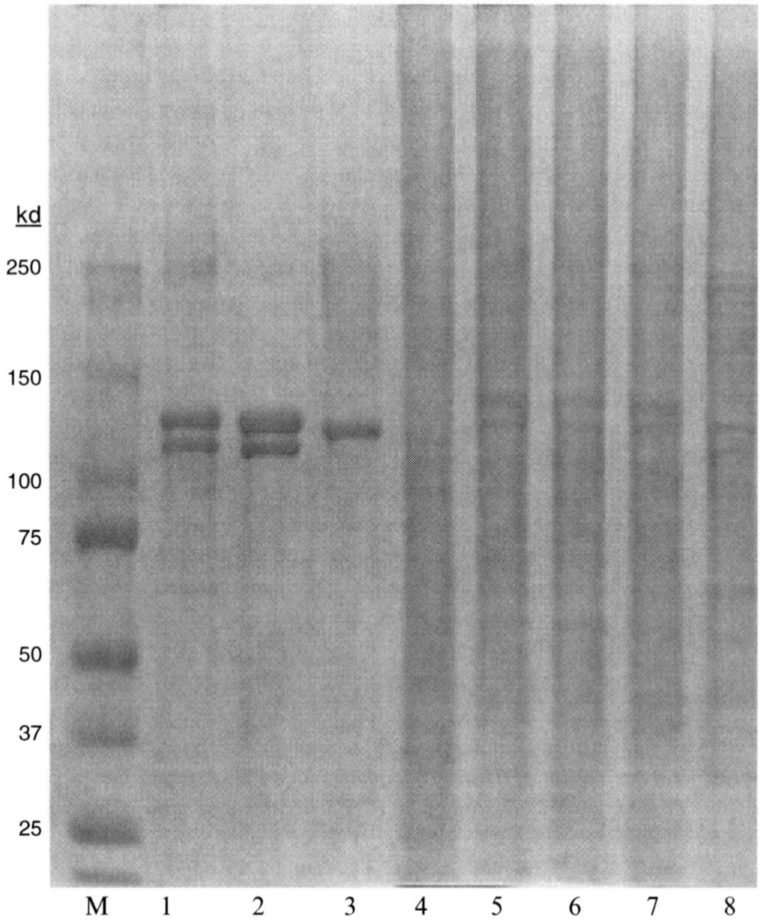

Fig. 7 SDS-PAGE analysis of capsule gelatins and FG-5020. M, BioRad Precision Plus molecular weight marker; lane 1–3 μg human placental type I collagen; lane 2–3 μg recombinant human type I collagen; lane 3–5 μg of FG-5012; lane 4–10 μg type A gelatin, Bloom 195 from porcine bone, lane 5–10 μg type B gelatin, Bloom 150 from porcine skin, lane 6–10 μg type B gelatin blend, Bloom 150/190 from bone, lane 7–10 μg type B gelatin, Bloom 175 from bovine hide; lane 8–10 μg blend of type A and type B gelatin from porcine bone and skin

same range as several commercially available gelatins that are used in the food, nutracuetical, photographic and pharmaceutical industry.

In certain applications, an important performance criterion for gelatin is the ability to maintain its viscosity when incubated at elevated temperature for significant periods of time (thermoviscosity breakdown). This criterion was monitored by testing viscosity before and after a 17 hour incubation at 60°C. Typically, a viscosity loss of about 20% is observed with animal gelatins (Reich 2004). The viscosity of FG-5020 decreased 3.5% following the thermal stress. The viscosity of porcine type A gelatin from bone or hide and bovine type B gelatin decreased 20–30% over the course of the experiment. The results of this comparison are summarized in

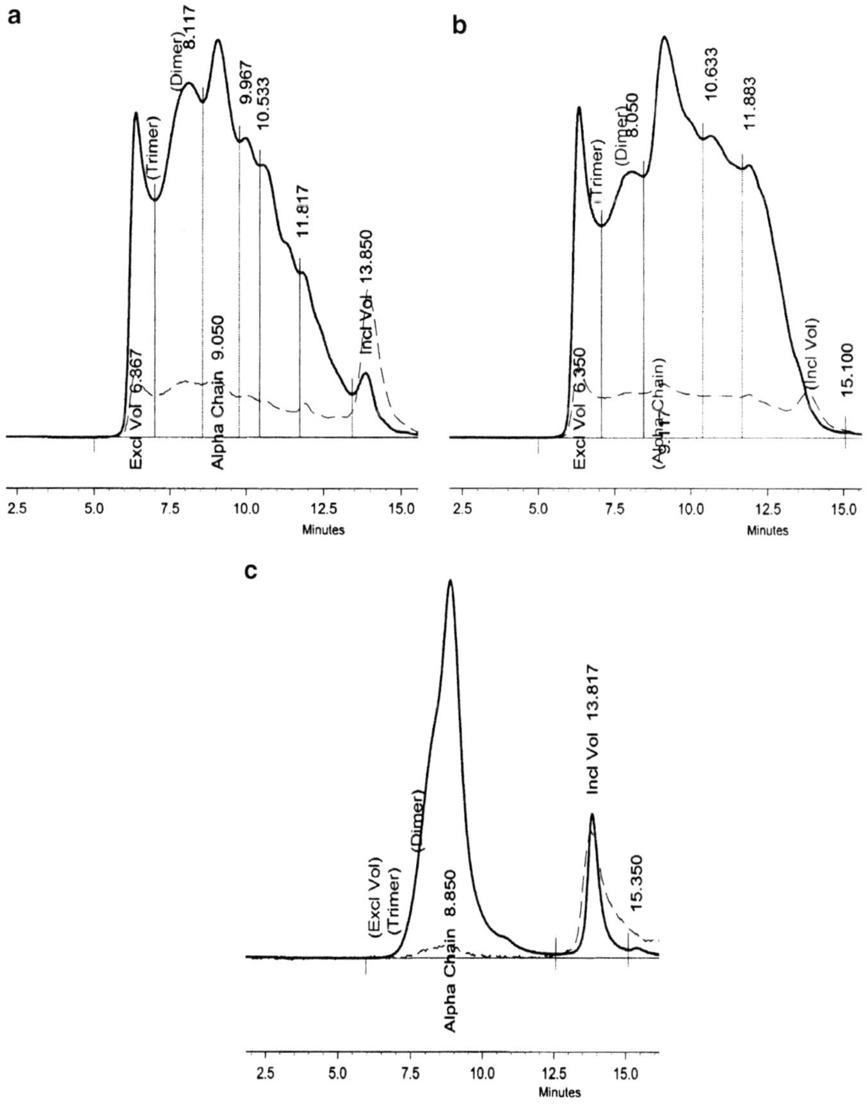

Fig. 8 Analysis of molecular weight distribution by size exclusion chromatography. Approximately 50 µg of type A (porcine hide, *Panel A*), type B (bovine hide, *Panel B*), or FG-5020 (*Panel C*) were analyzed by size exclusion chromatography on a Bio-Sil SEC 400–5 column (300 × 7.8 mm) in 2 M guanidine hydrochloride. Absorbance was monitored 215 nm

Table 4. This result demonstrates FG-5020 loses significantly less viscosity and indicates the recombinant material should perform as well or better than animal derived gelatin when maintenance of viscosity is an important process parameter.

As mentioned above, cross-linking of gelatin in capsules can be problematic. Lysine residues in the gelatin backbone have been identified as sites of cross-link formation. We are developing an engineered gelatin which lacks lysine residues and

Table 3 Bloom and viscosity measurements for FG-5020 and animal gelatin

Preparation	Bloom strength[a] (g)	Viscosity[b] (cP)
FG-5020	225	3.69
Bovine Type B gelatin	225	4.0

[a]Bloom strength was measured using a 6.6% solution of gelatin at 10°C.

[b]Viscosity was measured using a 6.6% solution of gelatin at 60°C.

Table 4 Comparison of viscosity breakdown at 60°C

Preparation (20% solution)	Viscosity (cP) T_0	$T_{17\ h}$	Percent decrease
FG-5020	42.5	41.0	3.5
Porcine Type A (175 Bloom)	35.3	27.7	21.5
Porcine Type A (300 Bloom)	102.5	71.0	30.7
Bovine Type B (225 Bloom)	41.1	31.2	24.1

therefore should not be as susceptible to cross-link formation; thus it may exhibit superior performance properties. The ability to manipulate the gene sequence expressed in *Pichia* exemplifies the advantage of this recombinant system over tissue derived gelatins and demonstrates our ability to produce designer gelatins with enhanced performance features.

Analysis of Gelatin Structure

Collagen, the protein from which gelatins are derived, has a characteristic circular dichroism (CD) profile where the far-ultraviolet spectrum shows a negative ellipticity around 197–199 nm and maximum in the 220–225 nm range. This profile indicates the presence of a left-handed polyproline II secondary structure. Following the thermal denaturation of collagen (conversion to gelatin) the positive CD peak at 220–225 nm is lost, indicative of the loss of the triple helical structure (Mizuno et al. 2003).

Animal gelatins, which form three-dimensional gels following a heating and cooling cycle, are thought to regain some amount of triple helical structure but a complete, intact triple helix is not reformed. To determine whether our recombinant gelatins have the ability to form triple helices we analyzed them by CD at 20°C. Type I collagen was analyzed in parallel as a positive control. The spectra from the 8.5, 25, 50, and 100 kDa gelatins all looked very similar. The typical negative ellipticity at 197–199 nm was observed but no maxima at 220–225 nm was seen (Fig. 9). The spectra from the type I collagen control yielded the typical pattern. These data indicate the recombinant gelatins do not form triple helices under the conditions used for analysis.

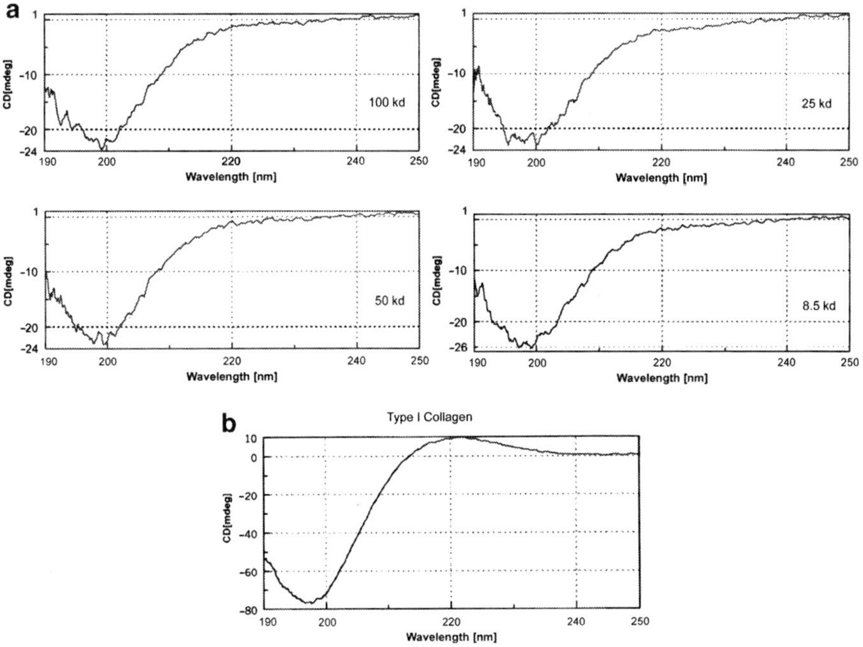

Fig. 9 Analysis of recombinant gelatin structure by circular dichroism spectropolarimetry. Samples were dissolved in 10 mm HCl at 125 μg/mL and analyzed on a Jasco model J-715 spectropoloarimeter with a peltier controlled sample holder. samples were scanned at wavelengths of 190 to 250 nm. The scans from the 8.5, 25, 50, and 100 kDa gelatins are shown in *panel a* and the same analysis performed on triple helical type I collagen is shown in *Panel b*

Summary

We have developed an expression system for the production of recombinant human gelatins ranging in size from 56 to 1,014 amino acids. These gelatins are expressed and secreted in the extracellular broth by the yeast *Pichia pastoris*. We have exemplified the utility of protein engineering to enhance the quality of these recombinant gelatins. These gelatins have defined molecular weights and pI and can be manufactured using GMPs. We have demonstrated their utility as stabilizers of biologics, as plasma expanders and we have shown that a 100 kDa gelatin has the required attributes to replace the animal gelatins that form solid gels. These recombinant gelatins represent a highly purified, reproducible, and safe alternative to the animal-derived gelatins that are widely used in the pharmaceutical industry.

Acknowledgements The authors would like to thank Elaine Lee for preparation of this manuscript. Table 2 is reprinted from *Advanced Drug Delivery Reviews 55:1547–1567, 2003; Olsen D.* et al., *Recombinant Collagen and Gelatin for Drug Delivery* with permission from Elsevier.

References

Boehm T, Pirie-Shepherd S, Trinh LB, Shiloach J, Folkman J (1999) Disruption of the KEX1 gene in *Pichia pastoris* allows expression of full-length murine and human endostatin. Yeast 15:563–572

Brierley RA (1998) Secretion of recombinant human insulin-like growth factor I (IGF-I). Meth Mol Biol 103:149–177

Bruckner P, Prockop DJ (1981) Proteolytic enzymes as probes for the triple-helical conformation of procollagen. Anal Biochem 110:360–368

Burke CJ, Hsu TA, Volkin DB (1999) Formulation, stability, and delivery of live attenuated vaccines for human use. Crit Rev Ther Drug Carrier Syst 16:1–83

Carstensen J, Rhodes C (1993) Pellicle formation in gelatin capsules. Drug Dev Ind Pharm 19:2709–2712

Clare JJ, Romanos MA, Rayment FB, Rowedder JE, Smith MA, Payne MM et al (1991) Production of mouse epidermal growth factor in yeast: high-level secretion using *Pichia pastoris* strains containing multiple gene copies. Gene 105:205–212

Cregg JM, Vedvick TS, Raschke WC (1993) Recent advances in the expression of foreign genes in *Pichia pastoris*. Biotechnology 11:905–910

deRizzo E, Tenorio E, Mendes I, Fang F, Pral M, Takata C et al (1989) Sorbitol-gelatin and glutamic-lactose solutions for stabilization of reference preparations of measles virus. Bull Pan Am Health Organ 23:299–305

Gholap D, Singh S (2004) Tje influence of drugs on gelatin cross-linking. PharmTechnol 94–102

Gold TB, Smith SL, Digenis GA (1996) Studies on the influence of pH and pancreatin on 13C-formaldehyde-induced gelatin cross-links using nuclear magnetic resonance. Pharm Dev Technol 1:21–26

Goodrick JC, Xu M, Finnegan R, Schilling BM, Schiavi S, Hoppe H et al (2001) High-level expression and stabilization of recombinant human chitinase produced in a continuous constitutive *Pichia pastoris* expression system. Biotechnol Bioeng 74:492–497

Mhatre R, Malinowski H, Nguyen H, Meyer M, Straughn A, Lesko L et al (1997) The effects of crosslinking in gelatin capsules on the bioequivalence of acetaminophen. Pharm Res 14:S251

Mizuno K, Hayashi T, Bachinger HP (2003) Hydroxylation-induced stabilization of the collagen triple helix. Further characterization of peptides with 4(R)-hydroxyproline in the Xaa position. J Biol Chem 278:32373–32379

Montagnon B, Vincent-Falquet JC, Fanget B (1983) Thousand litre scale microcarrier culture of Vero cells for killed polio virus vaccine. Promising results. Dev Biol Stand 55:37–42

Ofner CM, Bubnis WA (1996) Chemical and swelling evaluations of amino group crosslinking in gelatin and modified gelatin matrices. Pharma Res 13:1821–1827

Ofner CM III, Zhang YE, Jobeck VC, Bowman BJ (2001) Crosslinking studies in gelatin capsules treated with formaldehyde and in capsules exposed to elevated temperature and humidity. J Pharm Sci 90:79–88

Olsen D, Yang C, Bodo M, Chang R, Leigh S, Baez J et al (2003) Recombinant collagen and gelatin for drug delivery. Adv Drug Deliv Rev 55:1547–1567

Olsen D, Jiang J, Chang R, Duffy R, Sakaguchi M, Leigh S et al (2005) Expression and characterization of a low molecular weight human gelatin. Development of a substitute for animal-derived gelatin with superior features. Protein Exp Purif

Reich G (2004) Formulation and physical properties of soft capsules. In: Podczeck F, Jones B (eds) Pharmaceutical capsules, 2nd edn. Pharmaceutical Press, London, pp 201–212

Saddler JM, Horsey PJ (1987) The new generation gelatins. A review of their history, manufacture and properties. Anaesthesia 42:998–1004

Sakaguchi M, Inouye S (2000) Systemic allergic reactions to gelatin included in vaccines as a stabilizer. Jpn J Infect Dis 53:189–195

Sakaguchi M, Ogura H, Inouye S (1995) IgE antibody to gelatin in children with immediate-type reactions to measles and mumps vaccines. J Allergy Clin Immunol 96:563–565

Sakaguchi M, Hori H, Hattori S, Irie S, Imai A, Yanagida M et al (1999) IgE reactivity to alpha1 and alpha2 chains of bovine type 1 collagen in children with bovine gelatin allergy. J Allergy Clin Immunol 104:695–699

Sakai Y, Yamato R, Onuma M, Kikuta T, Watanabe M, Nakayama T (1998) Non-antigenic and low allergic gelatin produced by specific digestion with an enzyme-coupled matrix. Biol Pharm Bull 21:330–334

Tourtellotte D, Williams H (1958) Acylated gelatins and their preparations. [2827419]

Vassileva A, Chugh DA, Swaminathan S, Khanna N (2001) Expression of hepatitis B surface antigen in the methylotrophic yeast *Pichia pastoris* using the GAP promoter. J Biotechnol 88:21–35

Vervloet D, Senft M, Dugue P, Arnaud A, Charpin J (1983) Anaphylactic reactions to modified fluid gelatins. J Allergy Clin Immunol 71:535–540

Werten MW, van den Bosch TJ, Wind RD, Mooibroek H, de Wolf F (1999) High-yield secretion of recombinant gelatins by *Pichia pastoris*. Yeast 15:1087–1096

Werten MWT, Wisselink WH, Jansen-van den Bosch T, de Bruin EC, de Wolf F (2001) Secreted production of a custom-designed, highly hydrophilic gelatin in *Pichia pastoris*. Protein Eng 14:447–454

White CE, Hunter MJ, Meininger DP, White LR, Komives EA (1995) Large-scale expression, purification and characterization of small fragments of thrombomodulin: the roles of the sixth domain and of methionine 388. Protein Eng 8:1177–1187

Zhang AL, Luo JX, Zhang TY, Chen SC, Guan WJ (2004) Constitutive expression of human angiostatin in *Pichia pastoris* using the GAP promoter. Yi Chuan Xue Bao 31:552–557

Index

CPSIA information can be obtained at www.ICGtesting.com
Printed in the USA
LVOW070921040412

276079LV00003B/11/P

9 789048 118939